Photosynthesis: Concepts and Applications

Photosynthesis: Concepts and Applications

Edited by
Jonathon Saunders

 Larsen & Keller
www.larsen-keller.com

Photosynthesis: Concepts and Applications
Edited by Jonathon Saunders
ISBN: 978-1-63549-220-0 (Hardback)

© 2017 Larsen & Keller

 Larsen & Keller

Published by Larsen and Keller Education,
5 Penn Plaza,
19th Floor,
New York, NY 10001, USA

Cataloging-in-Publication Data

Photosynthesis : concepts and applications / edited by Jonathon Saunders.
    p. cm.
Includes bibliographical references and index.
ISBN 978-1-63549-220-0
1. Photosynthesis. 2. Photobiology. 3. Plants--Effect of light on.
I. Saunders, Jonathon.
QK882 .P46 2017
571.2--dc23

The publisher's policy is to use permanent paper from mills that operate a sustainable forestry policy. Furthermore, the publisher ensures that the text paper and cover boards used have met acceptable environmental accreditation standards.

Printed and bound in the United States of America.

For more information regarding Larsen and Keller Education and its products, please visit the publisher's website www.larsen-keller.com

Table of Contents

Preface

This book elucidates the concepts and innovative models around prospective developments with respect to photosynthesis. It provides detailed information about its methods and uses. Photosynthesis refers to the process by which plants convert light energy into chemical energy. It is an essential process as it enables plants to grow and bear fruit. The aim of the text is to familarize the readers with the process and associated chemicals. Most of the topics introduced in this textbook are of utmost importance to gain understanding about the topic. Coherent flow of topics, student-friendly language and extensive use of examples make this text an invaluable source of knowledge. The topics covered in this book offer the readers new insights in the field of photosynthesis.

A foreword of all chapters of the book is provided below:

Chapter 1 - Most plants and certain organisms produce energy by the process of photosynthesis, where light energy is converted to chemical forms of energy. Photosynthesis accounts for the food we eat and the fossil fuel found on the Earth. This chapter provides the reader with an overview of the process of photosynthesis and talks about the crucial role it plays in the continuation of life on Earth; **Chapter 2** - The process of photosynthesis depends on the pigment that provides the green color - chlorophyll, which helps in the absorption of sunlight and certain other pigments like carotene, xanthophyll, pheophytin, etc. that work together as light-harvesting complexes. This chapter studies each of these pigments in detail with emphasis on their chemical structures and the role they play in the process of photosynthesis; **Chapter 3** - As the process of photosynthesis is highly complex, several specialized enzymes exist to facilitate it. This chapter deals with the enzymes such as alternative oxidase, plastid terminal oxidase and transketolase. Each of these enzymes play crucial roles in the various cycles of photosynthesis and the reader is provided in-depth information about how these enzymes interact with other biomolecules in cellular reactions; **Chapter 4** - When chlorophyll absorbs light, it sets into motion a series of reduction-oxidation reactions that ultimately result in the production of energy in the form of ATP molecules. Photosynthetic efficiency refers to the fraction of light that is converted into energy. This chapter provides information on light-dependent reactions, light-independent reactions, C4 carbon fixation, photodissociation and oxygen evolution. The reader is able to better understand the relevance each of these reactions has on photosynthesis; **Chapter 5** - This chapter studies the various aspects of photosynthesis like photophosphorylation, photorespiration, photosynthate partitioning and photosynthetically active radiation. To understand these processes, the chapter introduces readers to concepts like chloroplast, thylakoid and PI curve. There is also a section dedicated to anoxygenic photosynthesis carried out by anaerobes like green sulphur bacteria, purple bacteria, acidobacteria and heliobacteria; **Chapter 6** - The chapter studies organisms that utilize the process of photosynthesis to produce energy. These organisms include phototrophs, algae, cyanobacteria, purple bacteria, green sulfur bacteria and heliobacteria. The content studies their ecological distribution, taxonomy and classification. Photosynthesis is best understood in confluence with the major topics listed in the following chapter; **Chapter 7** - Bio-inspired energy technology may one day be equipped with devices that can mimic the natural photosynthesis process and convert water into hydrogen ions and oxygen through the use of sunlight. Such technologies, when perfected, can provide immediately consumable energy which is completely environment friendly. This chapter will provide an integrated understanding of artificial photosynthesis; **Chapter 8** - The first photosynthetic organisms can be traced back to 2450 million years ago according to geological records. This chapter chronicles the evolutionary history of photosynthetic organisms and photosynthetic pathways. The reader is presented with information about the families of phototrophs and other photosynthetic organisms that exist on Earth. It provides a comprehensive overview of the evolution of photosynthesis.

At the end, I would like to thank all the people associated with this book devoting their precious time and providing their valuable contributions to this book. I would also like to express my gratitude to my fellow colleagues who encouraged me throughout the process.

Editor

Introduction to Photosynthesis

Most plants and certain organisms produce energy by the process of photosynthesis, where light energy is converted to chemical forms of energy. Photosynthesis accounts for the food we eat and the fossil fuel found on the Earth. This chapter provides the reader with an overview of the process of photosynthesis and talks about the crucial role it plays in the continuation of life on Earth.

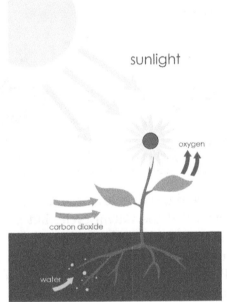

Schematic of photosynthesis in plants. The carbohydrates produced are stored in or used by the plant.

Photosynthesis is a process used by plants and other organisms to convert light energy, normally from the Sun, into chemical energy that can be later released to fuel the organisms' activities (energy transformation). This chemical energy is stored in carbohydrate molecules, such as sugars, which are synthesized from carbon dioxide and water. In most cases, oxygen is also released as a waste product. Most plants, most algae, and cyanobacteria perform photosynthesis; such organisms are called photoautotrophs. Photosynthesis maintains atmospheric oxygen levels and supplies all of the organic compounds and most of the energy necessary for life on Earth.

Although photosynthesis is performed differently by different species, the process always begins when energy from light is absorbed by proteins called reaction centres that contain green chlorophyll pigments. In plants, these proteins are held inside organelles called chloroplasts, which are most abundant in leaf cells, while in bacteria they are embedded in the plasma membrane. In these light-dependent reactions, some energy is used to strip electrons from suitable substances, such as water, producing oxygen gas. The hydrogen freed by water splitting is used in the creation of two further compounds: reduced nicotinamide adenine dinucleotide phosphate (NADPH) and adenosine triphosphate (ATP), the "energy currency" of cells.

$$6CO_2 + 6H_2O \xrightarrow{Light} C_6H_{12}O_6 + 6O_2$$

Carbon dioxide   Water   Sugar   Oxygen

Overall equation for the type of photosynthesis that occurs in plants

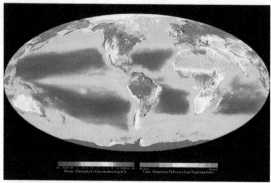

Composite image showing the global distribution of photosynthesis, including both oceanic phytoplankton and terrestrial vegetation. Dark red and blue-green indicate regions of high photosynthetic activity in ocean and land respectively.

In plants, algae and cyanobacteria, sugars are produced by a subsequent sequence of light-independent reactions called the Calvin cycle, but some bacteria use different mechanisms, such as the reverse Krebs cycle. In the Calvin cycle, atmospheric carbon dioxide is incorporated into already existing organic carbon compounds, such as ribulose bisphosphate (RuBP). Using the ATP and NADPH produced by the light-dependent reactions, the resulting compounds are then reduced and removed to form further carbohydrates, such as glucose.

The first photosynthetic organisms probably evolved early in the evolutionary history of life and most likely used reducing agents such as hydrogen or hydrogen sulfide, rather than water, as sources of electrons. Cyanobacteria appeared later; the excess oxygen they produced contributed to the oxygen catastrophe, which rendered the evolution of complex life possible. Today, the average rate of energy capture by photosynthesis globally is approximately 130 terawatts, which is about three times the current power consumption of human civilization. Photosynthetic organisms also convert around 100–115 thousand million metric tonnes of carbon into biomass per year.

Overview

Photosynthetic organisms are photoautotrophs, which means that they are able to synthesize food directly from carbon dioxide and water using energy from light. However, not all organisms that use light as a source of energy carry out photosynthesis, since *photoheterotrophs* use organic compounds, rather than carbon dioxide, as a source of carbon. In plants, algae and cyanobacteria, photosynthesis releases oxygen. This is called *oxygenic photosynthesis*. Although there are some differences between oxygenic photosynthesis in plants, algae, and cyanobacteria, the overall process is quite similar in these organisms. However, there are some types of bacteria that carry out anoxygenic photosynthesis. These consume carbon dioxide but do not release oxygen.

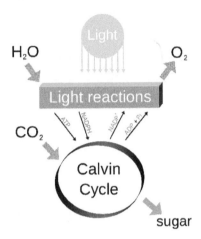

Photosynthesis changes sunlight into chemical energy, splits water to liberate O_2, and fixes CO_2 into sugar.

Carbon dioxide is converted into sugars in a process called carbon fixation. Carbon fixation is an endothermic redox reaction, so photosynthesis needs to supply both a source of energy to drive this process, and the electrons needed to convert carbon dioxide into a carbohydrate via a reduction reaction. The addition of electrons to a chemical species is called a reduction reaction. In general outline and in effect, photosynthesis is the opposite of cellular respiration, in which glucose and other compounds are oxidized to produce carbon dioxide and water, and to release chemical energy (an exothermic reaction) to drive the organism's metabolism. The two processes, of reduction of carbon dioxide to carbohydrate and then the later oxidation of the carbohydrate, take place through a different sequence of chemical reactions and in different cellular compartments.

The general equation for photosynthesis as first proposed by Cornelius van Niel is therefore:

CO_2 + 2H$_2$A + photons → [CH$_2$O] + 2A + H$_2$O carbon dioxide + electron donor + light energy → carbohydrate + oxidized electron donor + water

Since water is used as the electron donor in oxygenic photosynthesis, the equation for this process is:

CO_2 + 2H$_2$O + photons → [CH$_2$O] + O$_2$ + H$_2$O carbon dioxide + water + light energy → carbohydrate + oxygen + water

This equation emphasizes that water is both a reactant in the light-dependent reaction and a product of the light-independent reaction, but canceling n water molecules from each side gives the net equation:

$$CO_2 + H_2O + photons \rightarrow [CH_2O] + O_2 \text{ carbon dioxide + water + light energy} \rightarrow$$
carbohydrate + oxygen

Other processes substitute other compounds (such as arsenite) for water in the electron-supply role; for example some microbes use sunlight to oxidize arsenite to arsenate: The equation for this reaction is:

$$CO_2 + (AsO_3^{3-}) + photons \rightarrow (AsO_4^{3-}) + CO \text{ carbon dioxide + arsenite + light}$$
energy $\rightarrow$ arsenate + carbon monoxide (used to build other compounds in subsequent reactions)

Photosynthesis occurs in two stages. In the first stage, *light-dependent reactions* or *light reactions* capture the energy of light and use it to make the energy-storage molecules ATP and NADPH. During the second stage, the *light-independent reactions* use these products to capture and reduce carbon dioxide.

Most organisms that utilize photosynthesis to produce oxygen use visible light to do so, although at least three use shortwave infrared or, more specifically, far-red radiation.

Archaeobacteria use a simpler method using a pigment similar to the pigments used for vision. The archaearhodopsin changes its configuration in response to sunlight, acting as a proton pump. This produces a proton gradient more directly which is then converted to chemical energy. The process does not involve carbon dioxide fixation and does not release oxygen. It seems to have evolved separately.

Photosynthetic Membranes and Organelles

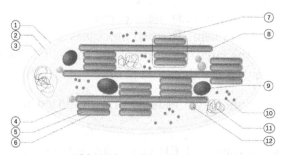

Chloroplast ultrastructure:
1. outer membrane, 2. intermembrane space, 3. inner membrane (1+2+3: envelope),
4. stroma (aqueous fluid), 5. thylakoid lumen (inside of thylakoid), 6. thylakoid membrane, 7. granum (stack of thylakoids), 8. thylakoid (lamella), 9. starch, 10. ribosome, 11. plastidial DNA, 12. plastoglobule (drop of lipids)

In photosynthetic bacteria, the proteins that gather light for photosynthesis are embedded in cell membranes. In its simplest form, this involves the membrane surrounding the cell itself. However, the membrane may be tightly folded into cylindrical sheets called thylakoids, or bunched up into round vesicles called *intracytoplasmic membranes*. These structures can fill most of the interior of a cell, giving the membrane a very large surface area and therefore increasing the amount of light that the bacteria can absorb.

In plants and algae, photosynthesis takes place in organelles called chloroplasts. A typical plant cell contains about 10 to 100 chloroplasts. The chloroplast is enclosed by a membrane. This membrane is composed of a phospholipid inner membrane, a phospholipid outer membrane, and an intermembrane space between them. Enclosed by the membrane is an aqueous fluid called the stroma. Embedded within the stroma are stacks of thylakoids (grana), which are the site of photosynthesis. The thylakoids appear as flattened disks. The thylakoid itself is enclosed by the thylakoid membrane, and within the enclosed volume is a lumen or thylakoid space. Embedded in the thylakoid membrane are integral and peripheral membrane protein complexes of the photosynthetic system, including the pigments that absorb light energy.

Plants absorb light primarily using the pigment chlorophyll. The green part of the light spectrum is not absorbed but is reflected which is the reason that most plants have a green color. Besides chlorophyll, plants also use pigments such as carotenes and xanthophylls. Algae also use chlorophyll, but various other pigments are present, such as phycocyanin, carotenes, and xanthophylls in green algae, phycoerythrin in red algae (rhodophytes) and fucoxanthin in brown algae and diatoms resulting in a wide variety of colors.

These pigments are embedded in plants and algae in complexes called antenna proteins. In such proteins, the pigments are arranged to work together. Such a combination of proteins is also called a light-harvesting complex.

Although all cells in the green parts of a plant have chloroplasts, the majority of those are found in specially adapted structures called leaves. Certain species adapted to conditions of strong sunlight and aridity, such as many Euphorbia and cactus species, have their main photosynthetic organs in their stems. The cells in the interior tissues of a leaf, called the mesophyll, can contain between 450,000 and 800,000 chloroplasts for every square millimeter of leaf. The surface of the leaf is coated with a water-resistant waxy cuticle that protects the leaf from excessive evaporation of water and decreases the absorption of ultraviolet or blue light to reduce heating. The transparent epidermis layer allows light to pass through to the palisade mesophyll cells where most of the photosynthesis takes place.

Light-dependent Reactions

In the light-dependent reactions, one molecule of the pigment chlorophyll absorbs one photon and loses one electron. This electron is passed to a modified form of chlorophyll called pheophytin, which passes the electron to a quinone molecule, starting the flow of electrons down an electron transport chain that leads to the ultimate reduction of NADP to NADPH. In addition, this creates a proton gradient (energy gradient) across the chloroplast membrane, which is used by ATP synthase in the synthesis of ATP. The chlorophyll molecule ultimately regains the electron it lost when a water molecule is split in a process called photolysis, which releases a dioxygen (O_2) molecule as a waste product.

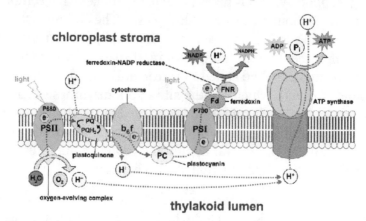

Light-dependent reactions of photosynthesis at the thylakoid membrane

The overall equation for the light-dependent reactions under the conditions of non-cyclic electron flow in green plants is:

$$2\ H_2O + 2\ NADP^+ + 3\ ADP + 3\ P_i + light \rightarrow 2\ NADPH + 2\ H^+ + 3\ ATP + O_2$$

Not all wavelengths of light can support photosynthesis. The photosynthetic action spectrum depends on the type of accessory pigments present. For example, in green plants, the action spectrum resembles the absorption spectrum for chlorophylls and carotenoids with peaks for violet-blue and red light. In red algae, the action spectrum is blue-green light, which allows these algae to use the blue end of the spectrum to grow in the deeper waters that filter out the longer wavelengths (red light) used by above ground green plants. The non-absorbed part of the light spectrum is what gives photosynthetic organisms their color (e.g., green plants, red algae, purple bacteria) and is the least effective for photosynthesis in the respective organisms.

Z Scheme

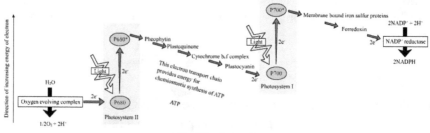

The "Z scheme"

In plants, light-dependent reactions occur in the thylakoid membranes of the chloroplasts where they drive the synthesis of ATP and NADPH. The light-dependent reactions are of two forms: cyclic and non-cyclic.

In the non-cyclic reaction, the photons are captured in the light-harvesting antenna complexes of photosystem II by chlorophyll and other accessory pigments. The absorption of a photon by the antenna complex frees an electron by a process called photoinduced charge separation. The antenna system is at the core of the chlorophyll molecule of the photosystem II reaction center. That freed electron is transferred to the primary electron-acceptor molecule, pheophytin. As the electrons are shuttled through an electron transport chain (the so-called *Z-scheme* shown in the diagram), it initially functions to generate a chemiosmotic potential by pumping proton cations (H^+) across the membrane and into the thylakoid space. An ATP synthase enzyme uses that chemiosmotic potential to make ATP during photophosphorylation, whereas NADPH is a product of the terminal redox reaction in the *Z-scheme*. The electron enters a chlorophyll molecule in Photosystem I. There it is further excited by the light absorbed by that photosystem. The electron is then passed along a chain of electron acceptors to which it transfers some of its energy. The energy delivered to the electron acceptors is used to move hydrogen ions across the thylakoid membrane into the lumen. The electron is eventually used to reduce the co-enzyme NADP with a H^+ to NADPH (which has functions in the light-independent reaction); at that point, the path of that electron ends.

The cyclic reaction is similar to that of the non-cyclic, but differs in that it generates only ATP, and no reduced NADP (NADPH) is created. The cyclic reaction takes place only at photosystem I. Once the electron is displaced from the photosystem, the electron is passed down the electron acceptor molecules and returns to photosystem I, from where it was emitted, hence the name *cyclic reaction*.

Water Photolysis

The NADPH is the main reducing agent produced by chloroplasts, which then goes on to provide a source of energetic electrons in other cellular reactions. Its production leaves chlorophyll in photosystem I with a deficit of electrons (chlorophyll has been oxidized), which must be balanced by some other reducing agent that will supply the missing electron. The excited electrons lost from chlorophyll from photosystem I are supplied from the electron transport chain by plastocyanin. However, since photosystem II is the first step of the *Z-scheme*, an external source of electrons is required to reduce its oxidized chlorophyll a molecules. The source of electrons in green-plant and cyanobacterial photosynthesis is water. Two water molecules are oxidized by four successive charge-separation reactions by photosystem II to yield a molecule of diatomic oxygen and four hydrogen ions; the electrons yielded are transferred to a redox-active tyrosine residue that then reduces the oxidized chlorophyll *a* (called P680) that serves as the primary light-driven electron donor in the photosystem II reaction center. That photo receptor is in effect reset and is then able to repeat the absorption of another photon and the release of another photo-dissociated electron. The oxidation of water is catalyzed in photosystem II by a redox-active structure that contains four manganese ions and a calcium ion; this oxygen-evolving complex binds two water molecules and contains the four oxidizing equivalents that are used to drive the water-oxidizing reaction. Photosystem II is the only known biological enzyme that carries out this oxidation of water. The hydrogen ions released contribute to the transmembrane chemiosmotic potential that leads to ATP synthesis. Oxygen is a waste product of light-dependent reactions, but the majority of organisms on Earth use oxygen for cellular respiration, including photosynthetic organisms.

Light-independent Reactions

Calvin Cycle

In the light-independent (or "dark") reactions, the enzyme RuBisCO captures CO_2 from the atmosphere and, in a process called the Calvin-Benson cycle, it uses the newly formed NADPH and releases three-carbon sugars, which are later combined to form sucrose and starch. The overall equation for the light-independent reactions in green plants is

$3 \, CO_2 + 9 \, ATP + 6 \, NADPH + 6 \, H^+ \rightarrow C_3H_6O_3\text{-phosphate} + 9 \, ADP + 8 \, P_i + 6 \, NADP^+ + 3 \, H_2O$

Carbon fixation produces the intermediate three-carbon sugar product, which is then converted to the final carbohydrate products. The simple carbon sugars produced by photosynthesis are then used in the forming of other organic compounds, such as the building material cellulose, the precursors for lipid and amino acid biosynthesis, or as a fuel in cellular respiration. The latter occurs not only in plants but also in animals when the energy from plants is passed through a food chain.

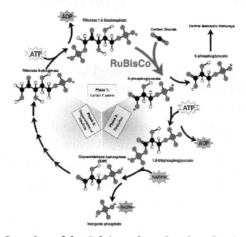

Overview of the Calvin cycle and carbon fixation

The fixation or reduction of carbon dioxide is a process in which carbon dioxide combines with a five-carbon sugar, ribulose 1,5-bisphosphate, to yield two molecules of a three-carbon compound, glycerate 3-phosphate, also known as 3-phosphoglycerate. Glycerate 3-phosphate, in the presence of ATP and NADPH produced during the light-dependent stages, is reduced to glyceraldehyde 3-phosphate. This product is also referred to as 3-phosphoglyceraldehyde (PGAL) or, more generically, as triose phosphate. Most (5 out of 6 molecules) of the glyceraldehyde 3-phosphate produced is used to regenerate ribulose 1,5-bisphosphate so the process can continue. The triose phosphates not thus "recycled" often condense to form hexose phosphates, which ultimately yield sucrose, starch and cellulose. The sugars produced during carbon metabolism yield carbon skeletons that can be used for other metabolic reactions like the production of amino acids and lipids.

Carbon Concentrating Mechanisms

On Land

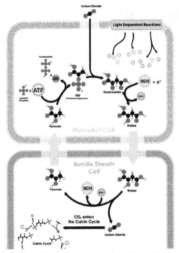

Overview of C4 carbon fixation

In hot and dry conditions, plants close their stomata to prevent water loss. Under these conditions, CO_2 will decrease and oxygen gas, produced by the light reactions of photosynthesis, will increase, causing an increase of photorespiration by the oxygenase activity of ribulose-1,5-bisphosphate carboxylase/oxygenase and decrease in carbon fixation. Some plants have evolved mechanisms to increase the CO_2 concentration in the leaves under these conditions.

Plants that use the C_4 carbon fixation process chemically fix carbon dioxide in the cells of the mesophyll by adding it to the three-carbon molecule phosphoenolpyruvate (PEP), a reaction catalyzed by an enzyme called PEP carboxylase, creating the four-carbon organic acid oxaloacetic acid. Oxaloacetic acid or malate synthesized by this process is then translocated to specialized bundle sheath cells where the enzyme RuBisCO and other Calvin cycle enzymes are located, and where CO_2 released by decarboxylation of the four-carbon acids is then fixed by RuBisCO activity to the three-carbon 3-phosphoglyceric acids. The physical separation of RuBisCO from the oxygen-generating light reactions reduces photorespiration and increases CO_2 fixation and, thus, the photosynthetic capacity of the leaf. C_4 plants can produce more sugar than C_3 plants in conditions of high light and temperature. Many important crop plants are C_4 plants, including maize, sorghum, sugarcane, and millet. Plants that do not use PEP-carboxylase in carbon fixation are called C_3 plants because the primary carboxylation reaction, catalyzed by RuBisCO, produces the three-carbon 3-phosphoglyceric acids directly in the Calvin-Benson cycle. Over 90% of plants use C_3 carbon fixation, compared to 3% that use C_4 carbon fixation; however, the evolution of C_4 in over 60 plant lineages makes it a striking example of convergent evolution.

Xerophytes, such as cacti and most succulents, also use PEP carboxylase to capture carbon dioxide in a process called Crassulacean acid metabolism (CAM). In contrast to C_4 metabolism, which *physically* separates the CO_2 fixation to PEP from the Calvin cycle, CAM *temporally* separates these two processes. CAM plants have a different leaf anatomy from C_3 plants, and fix the CO_2 at night, when their stomata are open. CAM plants store the CO_2 mostly in the form of malic acid via carboxylation of phosphoenolpyruvate to oxaloacetate, which is then reduced to malate. Decarboxylation of malate during the day releases CO_2 inside the leaves, thus allowing carbon fixation to 3-phosphoglycerate by RuBisCO. Sixteen thousand species of plants use CAM.

In Water

Cyanobacteria possess carboxysomes, which increase the concentration of CO_2 around RuBisCO to increase the rate of photosynthesis. An enzyme, carbonic anhydrase, located within the carboxysome releases CO_2 from the dissolved hydrocarbonate ions (HCO_3^-). Before the CO_2 diffuses out it is quickly sponged up by RuBisCO, which is concentrated within the carboxysomes. HCO_3^- ions are made from CO_2 outside the cell by another carbonic anhydrase and are actively pumped into the cell by a membrane protein. They cannot cross the membrane as they are charged, and within the cytosol they turn back into CO_2 very slowly without the help of carbonic anhydrase. This causes the HCO_3^- ions to accumulate within the cell from where they diffuse into the carboxysomes. Pyrenoids in algae and hornworts also act to concentrate CO_2 around rubisco.

Order and Kinetics

The overall process of photosynthesis takes place in four stages:

Stage	Description	Time scale
1	Energy transfer in antenna chlorophyll (thylakoid membranes)	femtosecond to picosecond
2	Transfer of electrons in photochemical reactions (thylakoid membranes)	picosecond to nanosecond
3	Electron transport chain and ATP synthesis (thylakoid membranes)	microsecond to millisecond
4	Carbon fixation and export of stable products	millisecond to second

Efficiency

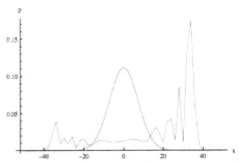

Probability distribution resulting from one-dimensional discrete time random walks. The quantum walk created using the Hadamard coin is plotted (blue) vs a classical walk (red) after 50 time steps.

Plants usually convert light into chemical energy with a photosynthetic efficiency of 3–6%. Absorbed light that is unconverted is dissipated primarily as heat, with a small fraction (1–2%) re-emitted as chlorophyll fluorescence at longer (redder) wavelengths.

Actual plants' photosynthetic efficiency varies with the frequency of the light being converted, light intensity, temperature and proportion of carbon dioxide in the atmosphere, and can vary from 0.1% to 8%. By comparison, solar panels convert light into electric energy at an efficiency of approximately 6–20% for mass-produced panels, and above 40% in laboratory devices.

Photosynthesis measurement systems are not designed to directly measure the amount of light absorbed by the leaf. But analysis of chlorophyll-fluorescence, P700- and P515-absorbance and gas exchange measurements reveal detailed information about e.g. the photosystems, quantum efficiency and the CO_2 assimilation rates. With some instruments even wavelength-dependency of the photosynthetic efficiency can be analyzed.

A phenomenon known as quantum walk increases the efficiency of the energy transport of light significantly. In the photosynthetic cell of an algae, bacterium, or plant, there are light-sensitive molecules called chromophores arranged in an antenna-shaped structure named a photocomplex. When a photon is absorbed by a chromophore, it is converted into a quasiparticle referred to as an exciton, which jumps from chromophore to chromophore towards the reaction center of the photocomplex, a collection of molecules that traps its energy in a chemical form that makes it accessible for the cell's metabolism. The exciton's wave properties enable it to cover a wider area and try out several possible paths simultaneously, allowing it to instantaneously "choose" the most efficient route, where it will have the highest probability of arriving at its destination in the minimum possible time. Because that quantum walking takes place at temperatures far higher than quantum phenomena usually occur, it is only possible over very short distances, due to obstacles in the form of destructive interference that come into play. These obstacles cause the particle to lose its wave properties for an instant before it regains them once again after it is freed from its locked position through a classic "hop". The movement of the electron towards the photo center is therefore covered in a series of conventional hops and quantum walks.

Early photosynthetic systems, such as those in green and purple sulfur and green and purple nonsulfur bacteria, are thought to have been anoxygenic, and used various other molecules as electron donors rather than water. Green and purple sulfur bacteria are thought to have used hydrogen and sulfur as electron donors. Green nonsulfur bacteria used various amino and other organic acids as an electron donor. Purple nonsulfur bacteria used a variety of nonspecific organic molecules. The use of these molecules is consistent with the geological evidence that Earth's early atmosphere was highly reducing at that time.

Fossils of what are thought to be filamentous photosynthetic organisms have been dated at 3.4 billion years old.

The main source of oxygen in the Earth's atmosphere derives from oxygenic photosynthesis, and its first appearance is sometimes referred to as the oxygen catastrophe. Geological evidence suggests that oxygenic photosynthesis, such as that in cyanobacteria, became important during the Paleoproterozoic era around 2 billion years ago. Modern photosynthesis in plants and most photosynthetic prokaryotes is oxygenic. Oxygenic photosynthesis uses water as an electron donor, which is oxidized to molecular oxygen (O_2) in the photosynthetic reaction center.

Symbiosis and the Origin of Chloroplasts

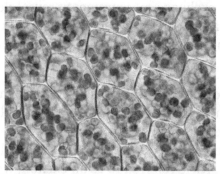

Plant cells with visible chloroplasts (from a moss, *Plagiomnium affine*)

Several groups of animals have formed symbiotic relationships with photosynthetic algae. These are most common in corals, sponges and sea anemones. It is presumed that this is due to the particularly simple body plans and large surface areas of these animals compared to their volumes. In addition, a few marine mollusks *Elysia viridis* and *Elysia chlorotica* also maintain a symbiotic relationship with chloroplasts they capture from the algae in their diet and then store in their bodies. This allows the mollusks to survive solely by photosynthesis for several months at a time. Some of the genes from the plant cell nucleus have even been transferred to the slugs, so that the chloroplasts can be supplied with proteins that they need to survive.

An even closer form of symbiosis may explain the origin of chloroplasts. Chloroplasts have many similarities with photosynthetic bacteria, including a circular chromosome, prokaryotic-type ribosome, and similar proteins in the photosynthetic reaction center. The endosymbiotic theory suggests that photosynthetic bacteria were acquired (by endocytosis) by early eukaryotic cells to form the first plant cells. Therefore, chloroplasts may be photosynthetic bacteria that adapted to life inside plant cells. Like mitochondria, chloroplasts possess their own DNA, separate from the nuclear DNA of their plant host cells and the genes in this chloroplast DNA resemble those found in cyanobacteria. DNA in chloroplasts codes for redox proteins such as those found in the photosynthetic reaction centers. The CoRR Hypothesis proposes that this Co-location is required for Redox Regulation.

Cyanobacteria and the Evolution of Photosynthesis

The biochemical capacity to use water as the source for electrons in photosynthesis evolved once, in a common ancestor of extant cyanobacteria. The geological record indicates that this transforming event took place early in Earth's history, at least 2450–2320 million years ago (Ma), and, it is speculated, much earlier. Because the Earth's atmosphere contained almost no oxygen during the estimated development of photosynthesis, it is believed that the first photosynthetic cyanobacteria did not generate oxygen. Available evidence from geobiological studies of Archean (>2500 Ma) sedimentary rocks indicates that life existed 3500 Ma, but the question of when oxygenic photosynthesis evolved is still unanswered. A clear paleontological window on cyanobacterial evolution opened about 2000 Ma, revealing an already-diverse biota of blue-green algae. Cyanobacteria remained the principal primary producers of oxygen throughout the Proterozoic Eon (2500–543 Ma), in part because the redox structure of the oceans favored photoautotrophs capable of nitrogen fixation. Green algae joined blue-green algae as the major primary producers of oxygen on continental shelves near the end of the Proterozoic, but it was only with the Mesozoic (251–65 Ma) radiations of dinoflagellates, coccolithophorids, and diatoms did the primary production of oxygen in marine shelf waters take modern form. Cyanobacteria remain critical to marine ecosystems as primary producers of oxygen in oceanic gyres, as agents of biological nitrogen fixation, and, in modified form, as the plastids of marine algae.

The Oriental hornet (*Vespa orientalis*) converts sunlight into electric power using a pigment called xanthopterin. This is the first evidence of a member of the animal kingdom engaging in photosynthesis.

Discovery

Although some of the steps in photosynthesis are still not completely understood, the overall photosynthetic equation has been known since the 19th century.

Jan van Helmont began the research of the process in the mid-17th century when he carefully measured the mass of the soil used by a plant and the mass of the plant as it grew. After noticing that the soil mass changed very little, he hypothesized that the mass of the growing plant must come from the water, the only substance he added to the potted plant. His hypothesis was partially accurate — much of the gained mass also comes from carbon dioxide as well as water. However, this was a signaling point to the idea that the bulk of a plant's biomass comes from the inputs of photosynthesis, not the soil itself.

Joseph Priestley, a chemist and minister, discovered that, when he isolated a volume of air under an inverted jar, and burned a candle in it, the candle would burn out very quickly, much before it ran out of wax. He further discovered that a mouse could similarly "injure" air. He then showed that the air that had been "injured" by the candle and the mouse could be restored by a plant.

In 1778, Jan Ingenhousz, repeated Priestley's experiments. He discovered that it was the influence of sunlight on the plant that could cause it to revive a mouse in a matter of hours.

In 1796, Jean Senebier, a Swiss pastor, botanist, and naturalist, demonstrated that green plants consume carbon dioxide and release oxygen under the influence of light. Soon afterward, Nicolas-Théodore de Saussure showed that the increase in mass of the plant as it grows could not be due only to uptake of CO_2 but also to the incorporation of water. Thus, the basic reaction by which photosynthesis is used to produce food (such as glucose) was outlined.

Cornelis Van Niel made key discoveries explaining the chemistry of photosynthesis. By studying purple sulfur bacteria and green bacteria he was the first to demonstrate that photosynthesis is a light-dependent redox reaction, in which hydrogen reduces carbon dioxide.

Robert Emerson discovered two light reactions by testing plant productivity using different wavelengths of light. With the red alone, the light reactions were suppressed. When blue and red were combined, the output was much more substantial. Thus, there were two photosystems, one absorbing up to 600 nm wavelengths, the other up to 700 nm. The former is known as PSII, the latter is PSI. PSI contains only chlorophyll "a", PSII contains primarily chlorophyll "a" with most of the available chlorophyll "b", among other pigment. These include phycobilins, which are the red and blue pigments of red and blue algae respectively, and fucoxanthol for brown algae and diatoms. The process is most productive when the absorption of quanta are equal in both the PSII and PSI, assuring that input energy from the antenna complex is divided between the PSI and PSII system, which in turn powers the photochemistry.

Robert Hill thought that a complex of reactions consisting of an intermediate to cytochrome b_6 (now a plastoquinone), another is from cytochrome f to a step in the carbohydrate-generating mechanisms. These are linked by plastoquinone, which does require energy to reduce cytochrome f for it is a sufficient reductant. Further experiments to prove that the oxygen developed during the photosynthesis of green plants came from water, were performed by Hill in 1937 and 1939. He showed that isolated chloroplasts give off oxygen in the presence of unnatural reducing agents like iron oxalate, ferricyanide or benzoquinone after exposure to light. The Hill reaction is as follows:

$$2\ H_2O + 2\ A + (\text{light, chloroplasts}) \rightarrow 2\ AH_2 + O_2$$

where A is the electron acceptor. Therefore, in light, the electron acceptor is reduced and oxygen is evolved.

Samuel Ruben and Martin Kamen used radioactive isotopes to determine that the oxygen liberated in photosynthesis came from the water.

Melvin Calvin works in his photosynthesis laboratory.

Melvin Calvin and Andrew Benson, along with James Bassham, elucidated the path of carbon assimilation (the photosynthetic carbon reduction cycle) in plants. The carbon reduction cycle is known as the Calvin cycle, which ignores the contribution of Bassham and Benson. Many scientists refer to the cycle as the Calvin-Benson Cycle, Benson-Calvin, and some even call it the Calvin-Benson-Bassham (or CBB) Cycle.

Nobel Prize-winning scientist Rudolph A. Marcus was able to discover the function and significance of the electron transport chain.

Otto Heinrich Warburg and Dean Burk discovered the I-quantum photosynthesis reaction that splits the CO_2, activated by the respiration.

Louis N.M. Duysens and Jan Amesz discovered that chlorophyll a will absorb one light, oxidize cytochrome f, chlorophyll a (and other pigments) will absorb another light, but will reduce this same oxidized cytochrome, stating the two light reactions are in series.

Development of the Concept

In 1893, Charles Reid Barnes proposed two terms, *photosyntax* and *photosynthesis*, for the biological process of *synthesis of complex carbon compounds out of carbonic acid, in the presence of chlorophyll, under the influence of light*. Over time, the term *photosynthesis* came into common usage as the term of choice. Later discovery of anoxygenic photosynthetic bacteria and photophosphorylation necessitated redefinition of the term.

Factors

The leaf is the primary site of photosynthesis in plants.

There are three main factors affecting photosynthesis and several corollary factors. The three main are:

- Light irradiance and wavelength

- Carbon dioxide concentration

- Temperature.

Light Intensity (Irradiance), Wavelength and Temperature

The process of photosynthesis provides the main input of free energy into the biosphere, and is one of four main ways in which radiation is important for plant life.

The radiation climate within plant communities is extremely variable, with both time and space.

In the early 20th century, Frederick Blackman and Gabrielle Matthaei investigated the effects of light intensity (irradiance) and temperature on the rate of carbon assimilation.

- At constant temperature, the rate of carbon assimilation varies with irradiance, increasing as the irradiance increases, but reaching a plateau at higher irradiance.

- At low irradiance, increasing the temperature has little influence on the rate of carbon assimilation. At constant high irradiance, the rate of carbon assimilation increases as the temperature is increased.

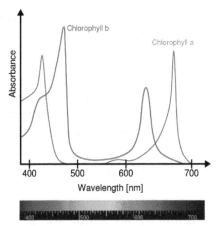

Absorbance spectra of free chlorophyll *a* (green) and *b* (red) in a solvent. The action spectra of chlorophyll molecules are slightly modified *in vivo* depending on specific pigment-protein interactions.

These two experiments illustrate several important points: First, it is known that, in general, photochemical reactions are not affected by temperature. However, these experiments clearly show that temperature affects the rate of carbon assimilation, so there must be two sets of reactions in the full process of carbon assimilation. These are, of course, the light-dependent 'photochemical' temperature-independent stage, and the light-independent, temperature-dependent stage. Second, Blackman's experiments illustrate the concept of limiting factors. Another limiting factor is the wavelength of light. Cyanobacteria, which reside several meters underwater, cannot receive the correct wavelengths required to cause photoinduced charge separation in conventional photosynthetic pigments. To combat this problem, a series of proteins with different pigments surround the reaction center. This unit is called a phycobilisome.

Carbon Dioxide Levels and Photorespiration

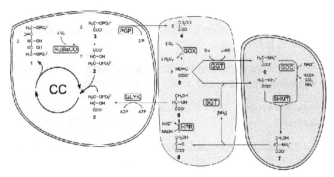

Photorespiration

As carbon dioxide concentrations rise, the rate at which sugars are made by the light-independent reactions increases until limited by other factors. RuBisCO, the enzyme that captures carbon dioxide in the light-independent reactions, has a binding affinity for both carbon dioxide and oxygen. When the concentration of carbon dioxide is high, RuBisCO will fix carbon dioxide. However, if the carbon dioxide concentration is low,

RuBisCO will bind oxygen instead of carbon dioxide. This process, called photorespiration, uses energy, but does not produce sugars.

RuBisCO oxygenase activity is disadvantageous to plants for several reasons:

1. One product of oxygenase activity is phosphoglycolate (2 carbon) instead of 3-phosphoglycerate (3 carbon). Phosphoglycolate cannot be metabolized by the Calvin-Benson cycle and represents carbon lost from the cycle. A high oxygenase activity, therefore, drains the sugars that are required to recycle ribulose 5-bisphosphate and for the continuation of the Calvin-Benson cycle.

2. Phosphoglycolate is quickly metabolized to glycolate that is toxic to a plant at a high concentration; it inhibits photosynthesis.

3. Salvaging glycolate is an energetically expensive process that uses the glycolate pathway, and only 75% of the carbon is returned to the Calvin-Benson cycle as 3-phosphoglycerate. The reactions also produce ammonia (NH_3), which is able to diffuse out of the plant, leading to a loss of nitrogen.

   A highly simplified summary is:

   2 glycolate + ATP $\rightarrow$ 3-phosphoglycerate + carbon dioxide + ADP + NH_3

The salvaging pathway for the products of RuBisCO oxygenase activity is more commonly known as photorespiration, since it is characterized by light-dependent oxygen consumption and the release of carbon dioxide.

References

- "Photosynthesis". McGraw-Hill Encyclopedia of Science & Technology. 13. New York: McGraw-Hill. 2007. ISBN 0-07-144143-3.

- Campbell NA, Williamson B, Heyden RJ (2006). Biology Exploring Life. Upper Saddle River, NJ: Pearson Prentice Hall. ISBN 0-13-250882-6.

- Raven PH, Evert RF, Eichhorn SE (2005). Biology of Plants, (7th ed.). New York: W.H. Freeman and Company Publishers. pp. 124–127. ISBN 0-7167-1007-2.

- Monson RK, Sage RF (1999). "16". C_4 plant biology. Boston: Academic Press. pp. 551–580. ISBN 0-12-614440-0.

- Herrero A, Flores E (2008). The Cyanobacteria: Molecular Biology, Genomics and Evolution (1st ed.). Caister Academic Press. ISBN 978-1-904455-15-8.

- Jones HG (2014). Plants and Microclimate: a Quantitative Approach to Environmental Plant Physiology (Third ed.). Cambridge: Cambridge University Press. ISBN 978-0-521-27959-8.

- Lloyd S (10 March 2014). "Quantum Biology: Better Living Through Quantum Mechanics - The Nature of Reality". Nova: PBS Online, WGBH Boston.

Photosynthetic Pigments: A Comprehensive Study

The process of photosynthesis depends on the pigment that provides the green color - chlorophyll, which helps in the absorption of sunlight and certain other pigments like carotene, xanthophyll, pheophytin, etc. that work together as light-harvesting complexes. This chapter studies each of these pigments in detail with emphasis on their chemical structures and the role they play in the process of photosynthesis.

Photosynthetic Pigment

A photosynthetic pigment (accessory pigment; chloroplast pigment; antenna pigment) is a pigment that is present in chloroplasts or photosynthetic bacteria and captures the light energy necessary for photosynthesis.

Plants

Green plants have six closely related photosynthetic pigments (in order of increasing polarity):

- Carotene - an orange pigment
- Xanthophyll - a yellow pigment
- Phaeophytin a - a gray-brown pigment
- Phaeophytin b - a yellow-brown pigment
- Chlorophyll a - a blue-green pigment
- Chlorophyll b - a yellow-green pigment

Chlorophyll a is the most common of the six, present in every plant that performs photosynthesis. The reason that there are so many pigments is that each absorbs light more efficiently in a different part of the electromagnetic spectrum. Chlorophyll a absorbs well at a wavelength of about 400-450 nm and at 650-700 nm; chlorophyll b at 450-500 nm and at 600-650 nm. Xanthophyll absorbs well at 400-530 nm. However, none of the pigments absorbs well in the green-yellow region, which is responsible for the abundant green we see in nature.

Bacteria

Like plants, the cyanobacteria use water as an electron donor for photosynthesis and therefore liberate oxygen; they also use chlorophyll as a pigment. In addition, most cyanobacteria use phycobiliproteins, water-soluble pigments which occur in the cytoplasm of the chloroplast, to capture light energy and pass it on to the chlorophylls. (Some cyanobacteria, the prochlorophytes, use chlorophyll b instead of phycobilin.) It is thought that the chloroplasts in plants and algae all evolved from cyanobacteria.

Several other groups of bacteria use the bacteriochlorophyll pigments (similar to the chlorophylls) for photosynthesis. Unlike the cyanobacteria, these bacteria do not produce oxygen; they typically use hydrogen sulfide rather than water as the electron donor.

Recently, a very different pigment has been found in some marine γ-proteobacteria: proteorhodopsin. It is similar to and probably originated from bacteriorhodopsin. Bacterial chlorophyll b has been isolated from Rhodopseudomonas spp. but its structure is not yet known

Algae

Green algae, red algae and glaucophytes all use chlorophylls. Red algae and glaucophytes also use phycobiliproteins, but green algae do not.

Archaea

Halobacteria use the pigment bacteriorhodopsin which acts directly as a proton pump when exposed to light.

Carotene

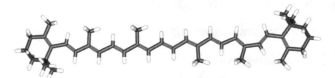

β-Carotene represented by a 3-dimensional stick diagram

The term carotene (also carotin, from the Latin *carota*, "carrot") is used for many related unsaturated hydrocarbon substances having the formula $C_{40}H_x$, which are synthesized by plants but in general cannot be made by animals (with the sole known exception of some aphids and spider mites which acquired the synthetic genes from fungi). Carotenes are photosynthetic pigments important for photosynthesis. Carotenes contain no oxygen atoms. They absorb ultraviolet, violet, and blue light and scatter orange or red light, and (in low concentrations) yellow light.

Carotene is responsible for the orange colour of carrots and the colours of many other fruits and vegetables and even some animals.

Lesser Flamingos in the Ngorongoro Crater, Tanzania. The pink colour of wild flamingos is due to astaxanthin (a carotenoid) they absorb from their diet of brine shrimp. If fed a carotene-free diet they become white.

Carotenes are responsible for the orange colour of the carrot, for which this class of chemicals is named, and for the colours of many other fruits, vegetables and fungi (for example, sweet potatoes, chanterelle and orange cantaloupe melon). Carotenes are also responsible for the orange (but not all of the yellow) colours in dry foliage. They also (in lower concentrations) impart the yellow coloration to milk-fat and butter. Omnivorous animal species which are relatively poor converters of coloured dietary carotenoids to colourless retinoids have yellowed-coloured body fat, as a result of the carotenoid retention from the vegetable portion of their diet. The typical yellow-coloured fat of humans and chickens is a result of fat storage of carotenes from their diets.

Carotenes contribute to photosynthesis by transmitting the light energy they absorb to chlorophyll. They also protect plant tissues by helping to absorb the energy from singlet oxygen, an excited form of the oxygen molecule O_2 which is formed during photosynthesis.

β-Carotene is composed of two retinyl groups, and is broken down in the mucosa of the human small intestine by β-carotene 15,15'-monooxygenase to retinal, a form of vitamin A. β-Carotene can be stored in the liver and body fat and converted to retinal as needed, thus making it a form of vitamin A for humans and some other mammals. The carotenes α-carotene and γ-carotene, due to their single retinyl group (β-ionone ring), also have some vitamin A activity (though less than β-carotene), as does the xanthophyll carotenoid β-cryptoxanthin. All other carotenoids, including lycopene, have no beta-ring and thus no vitamin A activity (although they may have antioxidant activity and thus biological activity in other ways).

Animal species differ greatly in their ability to convert retinyl (beta-ionone) containing carotenoids to retinals. Carnivores in general are poor converters of dietary ionone-containing carotenoids. Pure carnivores such as ferrets lack β-carotene 15,15'-monooxygenase and cannot convert any carotenoids to retinals at all (resulting in carotenes not being a form of vitamin A for this species); while cats can convert a trace of β-carotene to retinol, although the amount is totally insufficient for meeting their daily retinol needs.

Molecular Structure

Chemically, carotenes are polyunsaturated hydrocarbons containing 40 carbon atoms per molecule, variable numbers of hydrogen atoms, and no other elements. Some carotenes are terminated by hydrocarbon rings, on one or both ends of the molecule. All are coloured to the human eye, due to extensive systems of conjugated double bonds. Structurally carotenes are tetraterpenes, meaning that they are synthesized biochemically from four 10-carbon terpene units, which in turn are formed from eight 5-carbon isoprene units.

Carotenes are found in plants in two primary forms designated by characters from the Greek alphabet: alpha-carotene (α-carotene) and beta-carotene (β-carotene). Gamma-, delta-, epsilon-, and zeta-carotene (γ, δ, ε, and ζ-carotene) also exist. Since they are hydrocarbons, and therefore contain no oxygen, carotenes are fat-soluble and insoluble in water (in contrast with other carotenoids, the xanthophylls, which contain oxygen and thus are less chemically hydrophobic).

Dietary Sources

The following foods are particularly rich in carotenes :

- sweet potatoes
- carrots
- wolfberries (goji)
- cantaloupe melon
- mangoes
- apricots
- Persimmon
- spinach

- kale
- chard
- turnip greens
- dandelion greens
- beet greens
- mustard greens
- collard greens
- watercress
- cilantro (coriander)
- fresh thyme
- broccoli
- parsley
- romaine lettuce
- ivy gourd
- rose hips
- winter squash
- pumpkin
- cassava

Absorption from these foods is enhanced if eaten with fats, as carotenes are fat soluble, and if the food is cooked for a few minutes until the plant cell wall splits and the colour is released into any liquid. 6 µg of dietary β-carotene supplies the equivalent of 1 µg of retinol, or 1 RE (Retinol Equivalent). This is equivalent to 3⬚ IU of vitamin A.

The Multiple Forms

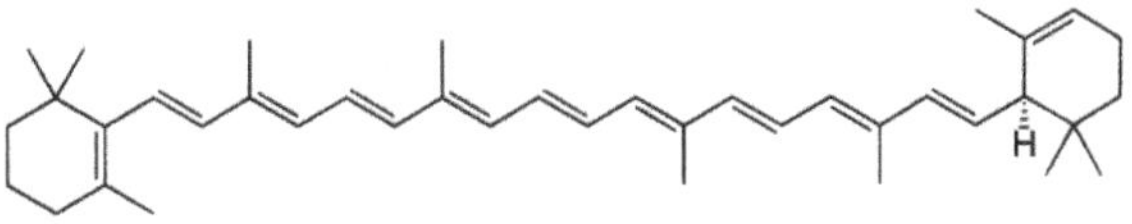

α-carotene

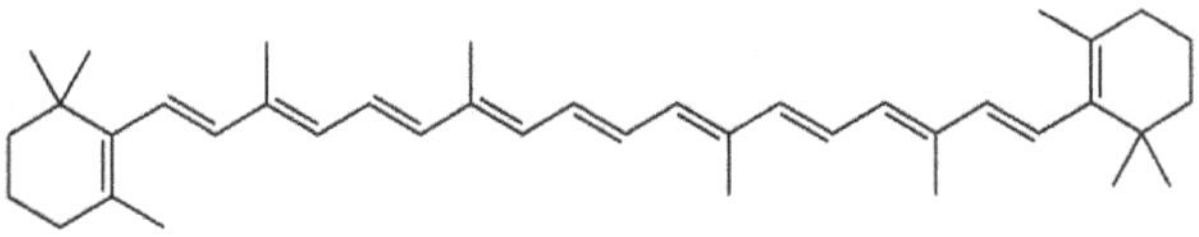

β-carotene

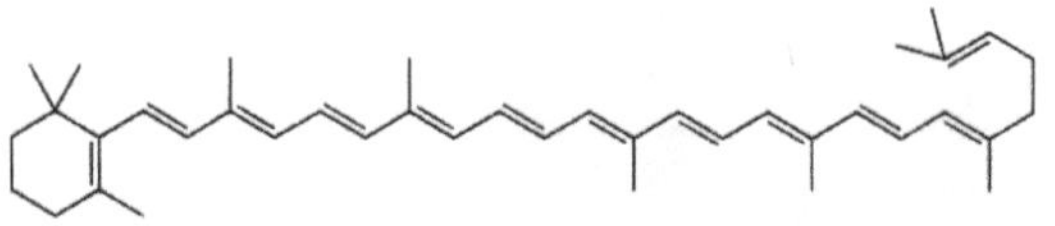

γ-carotene

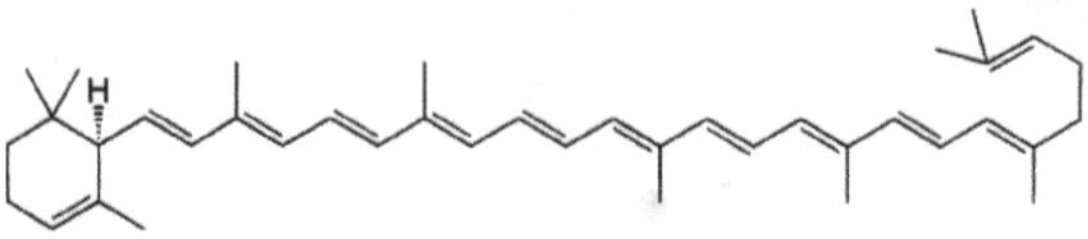

δ-carotene

The two primary isomers of carotene, α-carotene and β-carotene, differ in the position of a double bond (and thus a hydrogen) in the cyclic group at one end (the right end in the diagram at right).

β-Carotene is the more common form and can be found in yellow, orange, and green leafy fruits and vegetables. As a rule of thumb, the greater the intensity of the orange colour of the fruit or vegetable, the more β-carotene it contains.

Carotene protects plant cells against the destructive effects of ultraviolet light. β-Carotene is an antioxidant.

β-Carotene and Cancer

It has been shown in trials that the ingestion of β-carotene supplements at about 30 mg/day (10 times the Reference Daily Intake) increases the rate of lung and prostate cancer development in smokers and people with a history of asbestos exposure.

An article on the American Cancer Society says that The Cancer Research Campaign has called for warning labels on β-carotene supplements to caution smokers that such supplements may increase the risk of lung cancer.

The New England Journal of Medicine published an article in 1994 about a trial which examined the relationship between daily supplementation of β-carotene and vitamin E (α-tocopherol) and the incidence of lung cancer. The study was done using supplements and researchers were aware of the epidemiological correlation between carotenoid-rich fruits and vegetables and lower lung cancer rates. The research concluded that no reduction in lung cancer was found in the participants using these supplements, and furthermore, these supplements may, in fact, have harmful effects.

The Journal of the National Cancer Institute and The New England Journal of Medicine published articles in 1996 about a trial that was conducted to determine if vitamin A (in the form of retinyl palmitate) and β-carotene had any beneficial effects to prevent cancer. The results indicated an increased risk of lung cancer for the participants who consumed the β-carotene supplement and who had lung irritation from smoking or asbestos exposure, causing the trial to be stopped early.

A review of all randomized controlled trials in the scientific literature by the Cochrane Collaboration published in *JAMA* in 2007 found that synthetic β-carotene *increased* mortality by something between 1 and 8% (Relative Risk 1.05, 95% confidence interval 1.01–1.08). However, this meta-analysis included two large studies of smokers, so it is not clear that the results apply to the general population. The review only studied the influence of synthetic antioxidants and the results should not be translated to potential effects of fruits and vegetables.

β-Carotene and Cognition

A recent report demonstrated that 50 mg of β-carotene every other day prevented cognitive decline in a study of over 4000 physicians at a mean treatment duration of 18 years.

β-Carotene and Photosensitivity

Oral β-carotene is prescribed to people suffering from erythropoietic protoporphyria. It provides them some relief from photosensitivity.

β-Carotene and Nanotechnology

β-Carotene and lycopene molecules can be encapsulated into carbon nanotubes enhancing the optical properties of carbon nanotubes. Efficient energy transfer occurs between the encapsulated dye and nanotube — light is absorbed by the dye and without significant loss is transferred to the single wall carbon nanotube (SWCNT). Encapsulation increases chemical and thermal stability of carotene molecules; it also allows their isolation and individual characterization.

Carotenemia

Carotenemia or hypercarotenemia is excess carotene, but unlike excess vitamin A, carotene is non-toxic. Although hypercarotenemia is not particularly dangerous, it can lead to an oranging of the skin (carotenodermia), but not the conjunctiva of eyes (thus easily distinguishing it visually from jaundice). It is most commonly associated with consumption of an abundance of carrots, but it also can be a medical sign of more dangerous conditions.

Production

Most of the world's synthetic supply of carotene comes from a manufacturing complex located in Freeport, Texas and owned by DSM. The other major supplier BASF also uses a chemical process to produce β-carotene. Together these suppliers account for about 85% of the β-carotene on the market. In Spain Vitatene produces natural β-carotene from fungus Blakeslea trispora, as does DSM but at much lower amount when compared to its synthetic β-carotene operation. In Australia, organic β-carotene is produced by Aquacarotene Limited from dried marine algae *Dunaliella salina* grown in harvesting ponds situated in Karratha, Western Australia. BASF Australia is also producing β-carotene from microalgae grown in two sites in Australia that are the world's largest algae farms. In Portugal, the industrial biotechnology company Biotrend is producing natural all-*trans*-β-carotene from a non genetically modified bacteria of the *Sphingomonas* genus isolated from soil.

Algae farm ponds in Whyalla, South Australia, used to produce β-carotene.

Carotenes are also found in palm oil, corn, and in the milk of dairy cows, causing cow's milk to be light yellow, depending on the feed of the cattle, and the amount of fat in the milk (high-fat milks, such as those produced by Guernsey cows, tend to be yellower because their fat content causes them to contain more carotene).

Carotenes are also found in some species of termites, where they apparently have been picked up from the diet of the insects.

Total Synthesis

There are currently two commonly used methods of total synthesis of β-carotene. The first was developed by the Badische Anilin- & Soda-Fabrik (BASF) and is based on the Wittig reaction with Wittig himself as patent holder:

The second is a Grignard reaction, elaborated by Hoffman-La Roche from the original synthesis of Inhoffen et al. They are both symmetrical; the BASF synthesis is C20 + C20, and the Hoffman-La Roche synthesis is C19 + C2 + C19.

Nomenclature

Carotenes are carotenoids containing no oxygen. Carotenoids containing some oxygen are known as xanthophylls.

The two ends of the β-carotene molecule are structurally identical, and are called β-rings. Specifically, the group of nine carbon atoms at each end form a β-ring.

The α-carotene molecule has a β-ring at one end; the other end is called an ε-ring. There is no such thing as an "α-ring".

These and similar names for the ends of the carotenoid molecules form the basis of a systematic naming scheme, according to which:

- α-carotene is β,ε-carotene;

- β-carotene is β,β-carotene;

- γ-carotene (with one β ring and one uncyclized end that is labelled *psi*) is β,ψ-carotene;

- δ-carotene (with one ε ring and one uncyclized end) is ε,ψ-carotene;

- ε-carotene is ε,ε-carotene

- lycopene is ψ,ψ-carotene

ζ-Carotene is the biosynthetic precursor of neurosporene, which is the precursor of lycopene, which, in turn, is the precursor of the carotenes α through ε.

Food Additive

Carotene is also used as a substance to colour products such as juice, cakes, desserts, butter and margarine. It is approved for use as a food additive in the EU (listed as additive E160a) Australia and New Zealand (listed as 160a) and the US.

Xanthophyll

The color of an egg yolk is from the xanthophyll carotenoids lutein and zeaxanthin

Xanthophylls (originally phylloxanthins) are yellow pigments that occur widely in nature and form one of two major divisions of the carotenoid group; the other division is formed by the carotenes.

Molecular Structure

The molecular structure of xanthophylls is similar to that of carotenes, but xanthophylls contain oxygen atoms, while *carotenes* are purely hydrocarbons with no oxygen. Xanthophylls contain their oxygen either as hydroxyl groups and/or as pairs of hydrogen atoms that are substituted by oxygen atoms acting as a bridge (epoxide). For this reason, they are more polar than the purely hydrocarbon carotenes, and it is this difference that allows their separations from carotenes in many types of chromatography. Typically, carotenes are more orange in color than xanthophylls.

Occurrence

Like other carotenoids, xanthophylls are found in highest quantity in the leaves of most green plants, where they act to modulate light energy and perhaps serve as a non-photochemical quenching agent to deal with triplet chlorophyll (an excited form of chlorophyll), which is overproduced at high light levels in photosynthesis. The xanthophylls found in the bodies of animals, and in dietary animal products, are ultimately derived from plant sources in the diet. For example, the yellow color of chicken egg yolks, fat, and skin comes from ingested xanthophylls (primarily lutein, which is often added to chicken feed for this purpose).

The yellow color of the human macula lutea (literally, *yellow spot*) in the retina of the eye results from the lutein and zeaxanthin it contains, both xanthophylls again requiring a source in the human diet to be present in the eye. These xanthophylls protect the eye from ionizing blue and ultraviolet light, which they absorb. These two specific xanthophylls do not function in the mechanism of sight, since they cannot be converted to retinal (also called retinaldehyde or vitamin A aldehyde). Their arrangement is believed to be the cause of Haidinger's brush, an entoptic phenomenon that allows to perceive the polarization of light.

Example Compounds

The group of xanthophylls includes (among many other compounds) lutein, zeaxanthin, neoxanthin, violaxanthin, flavoxanthin, and α- and β-cryptoxanthin. The latter compound is the only known xanthophyll to contain a beta-ionone ring, and thus β-cryptoxanthin is the only xanthophyll that is known to possess pro-vitamin A activity for mammals. Even then, it is a vitamin only for plant-eating mammals that possess the enzyme to make retinal from carotenoids that contain beta-ionone (some carnivores lack this enzyme). In species other than mammals, certain xanthophylls may be converted to hydroxylated retinal-analogues that function directly in vision. For example, with the exception of certain flies, most insects use the xanthophyll derived R-isomer of 3-hydroxyretinal for visual activities, which means that β-cryptoxanthin and other xanthophylls (such as lutein and zeaxanthin) may function as forms of visual "vitamin A" for them, while carotenes (such as beta carotene) do not.

Xanthophyll Cycle

The xanthophyll cycle

The xanthophyll cycle involves the enzymatic removal of epoxy groups from xanthophylls (e.g. violaxanthin, antheraxanthin, diadinoxanthin) to create so-called de-epoxidised xanthophylls (e.g. diatoxanthin, zeaxanthin). These enzymatic cycles were found to play a key role in stimulating energy dissipation within light-harvesting antenna proteins by non-photochemical quenching- a mechanism to reduce the amount of energy that reaches the photosynthetic reaction centers. Non-photochemical quenching is one of the main ways of protecting against photoinhibition. In higher plants, there are three carotenoid pigments that are active in the xanthophyll cycle: violaxanthin, antheraxanthin, and zeaxanthin. During light stress, violaxanthin is converted to zeaxanthin via the intermediate antheraxanthin, which plays a direct photoprotective role acting as a lipid-protective anti-oxidant and by stimulating non-photochemical quenching within light-harvesting proteins. This conversion of violaxanthin to zeaxanthin is done by the enzyme violaxanthin de-epoxidase, while the reverse reaction is performed by zeaxanthin epoxidase.

In diatoms and dinoflagellates, the xanthophyll cycle consists of the pigment diadinoxanthin, which is transformed into diatoxanthin (diatoms) or dinoxanthin (dinoflagellates) under high-light conditions.

Wright et al. (Feb 2011) found that, "The increase in zeaxanthin appears to surpass the decrease in violaxanthin in spinach" and commented that the discrepancy could be explained by "a synthesis of zeaxanthin from beta-carotene", however they noted further study is required to explore this hypothesis.

Food Sources

Xanthophylls are found in all young leaves and in etiolated leaves. Examples of other rich sources include papaya, peaches, prunes, and squash, which contain lutein diesters.

Pheophytin

Pheophytin or phaeophytin (abbreviated Pheo) is a chemical compound that serves as the first electron carrier intermediate in the electron transfer pathway of photosystem II (PS II) in plants, and the photosynthetic reaction center (RC P870) found in purple bacteria. In both PS II and RC P870, light drives electrons from the reaction center through pheophytin, which then passes the electrons to a quinone (Q_A) in RC P870 and RC P680. The overall mechanisms, roles, and purposes of the pheophytin molecules in the two transport chains are analogous to each other.

Structure

In biochemical terms, pheophytin is a chlorophyll molecule lacking a central Mg ion. It can be produced from chlorophyll by treatment with a weak acid, producing a dark bluish waxy pigment. The probable etymology comes from this description, with *pheo* meaning *dusky* and *phyt* meaning *vegetation*.

History and Discovery

In the 1977s, scientists Klevanik, Klimov, Shuvalov performed a series of experiments to demonstrate that it is pheophytin and not plastoquinone that serves as the primary electron acceptor in photosystem II. Using several experiments, including electron paramagnetic resonance (EPR), they were able to show that pheophytin was reducible and, therefore, the primary electron acceptor between P680 and plastoquinone (Klimov, Allakhverdiev, Klevanik, Shuvalov). This discovery was met with fierce opposition, since many believed pheophytin to be a byproduct of chlorophyll degradation. Therefore, more experiments ensued to prove that pheophytin is indeed the primary electron acceptor of PSII, occurring between P680 and plastoquinone (Klimov, Allakhverdiev, Shuvalov). The data that was obtained is as follows:

1. Photo-reduction of pheophytin has been observed in various mixtures containing PSII reaction centers.

2. The quantity of pheophytin is in direct proportion to the number of PSII reaction centers.

3. Photo-reduction of pheophytin occurs at temperatures as low as 100K, and is observed after the reduction of plastoquinone.

These observations are all characteristic of photo-conversions of reaction center components.

Reaction in Purple Bacteria

Pheophytin is the first electron carrier intermediate in the photoreaction center (RC P870) of purple bacteria. Its involvement in this system can be broken down into 5 basic steps. The first step is excitation of the bacteriochlorophylls $(Chl)_2$ or the special pair of chlorophylls. This can be seen in the following reaction.

- $(Chl)_2 + 1$ photon $\rightarrow (Chl)_2^*$ (excitation)

The second step involves the $(Chl)_2$ passing an electron to pheophytin, producing a negatively charged radical (the pheophytin) and a positively charged radical (the special pair of chlorophylls), which results in a charge separation.

- $(Chl)_2^* + Pheo \rightarrow \cdot(Chl)_2^+ + \cdot Pheo^-$ (charge separation)

The third step is the rapid electron movement to the tightly bound menaquinone, Q_A, which immediately donates the electrons to a second, loosely bound quinine (Q_B). Two electron transfers convert Q_B to its reduced form (Q_BH_2).

- $2 \cdot Pheo^- + 2H^+ + Q_B \rightarrow 2Pheo + Q_BH_2$ (quinone reduction)

The fifth and final step involves the filling of the "hole" in the special pair by an electron from a heme in cytochrome c. This regenerates the substrates and completes the cycle, allowing for subsequent reactions to take place.

Involvement in Photosystem II

In photosystem II, pheophytin plays a very similar role. It again acts as the first electron carrier intermediate in the photosystem. After P680 becomes excited to P680*, it transfers an electron to pheophytin, which converts the molecule into a negatively charged radical. The negatively charged pheophytin radical quickly passes its extra electron to two consecutive plastoquinone molecules. Eventually, the electrons pass through the cytochrome b_6f molecule and leaves photosystem II. The reactions outlined above in the section concerning purple bacteria give a general illustration of the actual movement of the electrons through pheophytin and the photosystem. The overall scheme is:

1. Excitation

2. Charge separation

3. Plastoquinone reduction

4. Regeneration of substrates

Chlorophyll

Chlorophyll is responsible for the green color of many plants and algae.

There are several types of chlorophyll, but all share the chlorin magnesium ligand which forms the right-hand side of this diagram.

Chlorophyll (also chlorophyl) is a term used for several closely related green pigments found in cyanobacteria and the chloroplasts of algae and plants. Chlorophyll is an extremely important biomolecule, critical in photosynthesis, which allows plants to absorb energy from light. Chlorophyll absorbs light most strongly in the blue portion of the electromagnetic spectrum, followed by the red portion. Conversely, it is a poor absorber of green and near-green portions of the spectrum which it reflects, hence the green color of chlorophyll-containing tissues. Chlorophyll was first isolated and named by Joseph Bienaimé Caventou and Pierre Joseph Pelletier in 1817.

Chlorophyll and Photosynthesis

Chlorophyll is vital for photosynthesis, which allows plants to absorb energy from light.

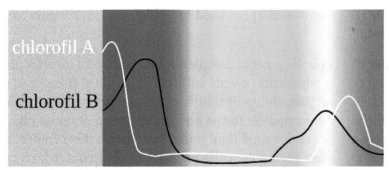

Absorption maxima of chlorophylls against the spectrum of white light. Chlorophyll absorbs energy from light strongly at visible frequencies.

Chlorophyll molecules are specifically arranged in and around photosystems that are embedded in the thylakoid membranes of chloroplasts. In these complexes, chlorophyll serves two primary functions. The function of the vast majority of chlorophyll (up to several hundred molecules per photosystem) is to absorb light and transfer that light energy by resonance energy transfer to a specific chlorophyll pair in the reaction center of the photosystems. The two currently accepted photosystem units are photosystem II and photosystem I, which have their own distinct reaction centres, named P680 and P700, respectively. These centres are named after the wavelength (in nanometers) of their red-peak absorption maximum. The identity, function and spectral properties of the types of chlorophyll in each photosystem are distinct and determined by each other and the protein structure surrounding them. Once extracted from the protein into a solvent (such as acetone or methanol), these chlorophyll pigments can be separated into chlorophyll a and chlorophyll b.

The function of the reaction center of chlorophyll is to absorb light energy and transfer it to other parts of the photosystem. The absorbed energy of the photon is transferred to an electron in a process called charge separation. The removal of the electron from the chlorophyll is an oxidation reaction. The chlorophyll donates the high energy electron to a series of molecular intermediates called an electron transport chain. The charged reaction center of chlorophyll ($P680^+$) is then reduced back to its ground state by accepting an electron stripped from water. The electron that reduces $P680^+$ ultimately comes from the oxidation of water into O_2 and H^+ through several intermediates. This reaction is how photosynthetic organisms such as plants produce O_2 gas, and is the source for practically all the O_2 in Earth's atmosphere. Photosystem I typically works in series with Photosystem II; thus the $P700^+$ of Photosystem I is usually reduced as it accepts the electron, via many intermediates in the thylakoid membrane, by electrons come, ultimately, from Photosystem II. Electron transfer reactions in the thylakoid membranes are complex, however, and the source of electrons used to reduce $P700^+$ can vary.

The electron flow produced by the reaction center chlorophyll pigments is used to pump H^+ ions across the thylakoid membrane, setting up a chemiosmotic potential used mainly in the production of ATP (stored chemical energy) or to reduce $NADP^+$ to NADPH. NADPH is a universal agent used to reduce CO_2 into sugars as well as other biosynthetic reactions.

Reaction center chlorophyll–protein complexes are capable of directly absorbing light and performing charge separation events without other the assistance of other chlorophyll pigments, but the probability of that happening under a given light intensity is small. Thus, the other chlorophylls in the photosystem and antenna pigment proteins all cooperatively absorb and funnel light energy to the reaction center. Besides chlorophyll a, there are other pigments, called accessory pigments, which occur in these pigment–protein antenna complexes.

Chemical Structure

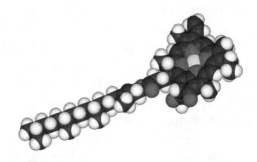

Space-filling model of the chlorophyll a molecule

Chlorophyll is a chlorin pigment, which is structurally similar to and produced through the same metabolic pathway as other porphyrin pigments such as heme. At the center of the chlorin ring is a magnesium ion. This was discovered in 1906, and was the first time that magnesium had been detected in living tissue. For the structures depicted in this article, some of the ligands attached to the Mg center are omitted for clarity. The chlorin ring can have several different side chains, usually including a long phytol chain. There are a few different forms that occur naturally, but the most widely distributed form in terrestrial plants is chlorophyll a. After initial work done by German chemist Richard Willstätter spanning from 1905 to 1915, the general structure of chlorophyll a was elucidated by Hans Fischer in 1940. By 1960, when most of the stereochemistry of chlorophyll a was known, Robert Burns Woodward published a total synthesis of the molecule. In 1967, the last remaining stereochemical elucidation was completed by Ian Fleming, and in 1990 Woodward and co-authors published an updated synthesis. Chlorophyll f was announced to be present in cyanobacteria and other oxygenic microorganisms that form stromatolites in 2010; a molecular formula of $C_{55}H_{70}O_6N_4Mg$ and a structure of (2-formyl)-chlorophyll a were deduced based on NMR, optical and mass spectra. The different structures of chlorophyll are summarized below:

	Chloro-phyll a	Chlorophyll b	Chlorophyll c1	Chloro-phyll c2	Chloro-phyll d	Chlorophyll f
Mo-lecular formula	$C_{55}H_{72}O_5N_4Mg$	$C_{55}H_{70}O_6N_4Mg$	$C_{35}H_{30}O_5N_4Mg$	$C_{35}H_{28}O_5N_4Mg$	$C_{54}H_{70}O_6N_4Mg$	$C_{55}H_{70}O_6N_4Mg$
C2 group	$-CH_3$	$-CH_3$	$-CH_3$	$-CH_3$	$-CH_3$	-CHO
C3 group	$-CH=CH_2$	$-CH=CH_2$	$-CH=CH_2$	$-CH=CH_2$	-CHO	$-CH=CH_2$
C7 group	$-CH_3$	-CHO	$-CH_3$	$-CH_3$	$-CH_3$	$-CH_3$
C8 group	$-CH_2CH_3$	$-CH_2CH_3$	$-CH_2CH_3$	$-CH=CH_2$	$-CH_2CH_3$	$-CH_2CH_3$
C17 group	$-CH_2CH_2COO$-Phytyl	$-CH_2CH_2COO$-Phytyl	$-CH=CHCOOH$	$-CH=CH-COOH$	$-CH_2CH_2COO$-Phytyl	$-CH_2CH_2COO$-Phytyl
C17-C18 bond	Single (chlorin)	Single (chlorin)	Double (porphyrin)	Double (porphy-rin)	Single (chlorin)	Single (chlorin)
Occur-rence	Universal	Mostly plants	Various algae	Various algae	Cyanobac-teria	Cyanobacteria

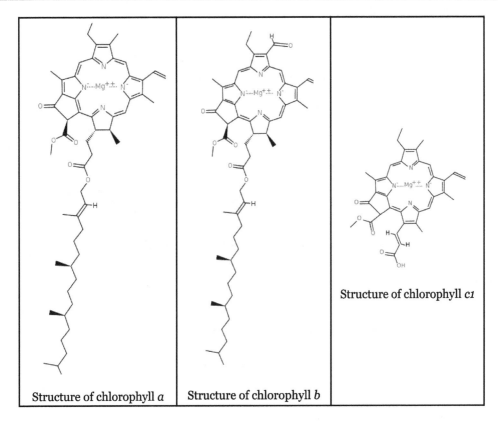

Structure of chlorophyll *a*   Structure of chlorophyll *b*

Structure of chlorophyll *c1*

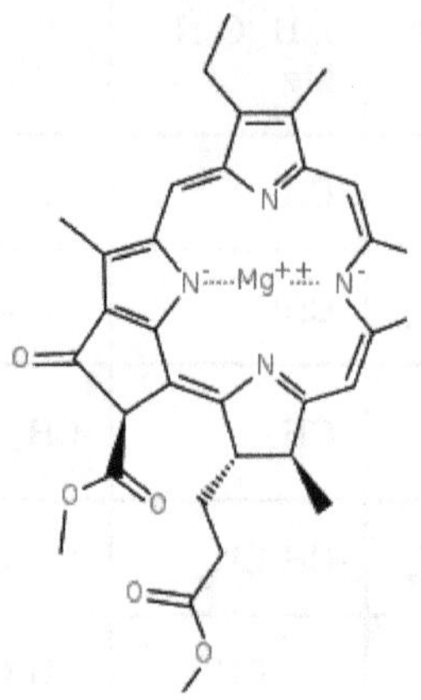

Structure of chlorophyll *c2*

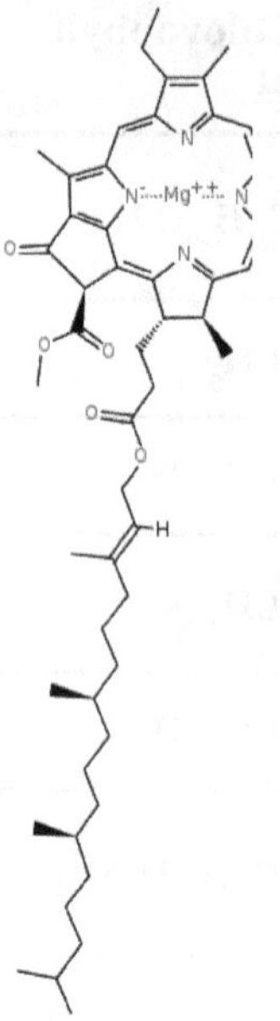

Structure of chlorophyll *d*

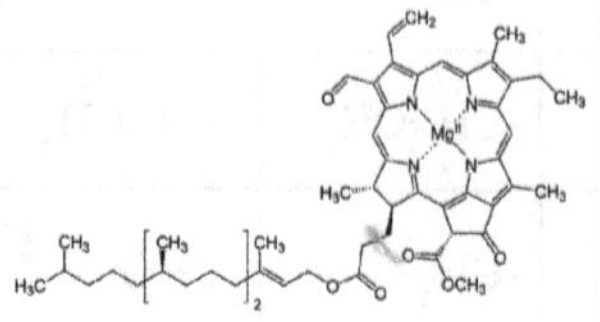

Structure of chlorophyll *f*

When leaves degreen in the process of plant senescence, chlorophyll is converted to a group of colourless tetrapyrroles known as nonfluorescent chlorophyll catabolites (NCC's) with the general structure:

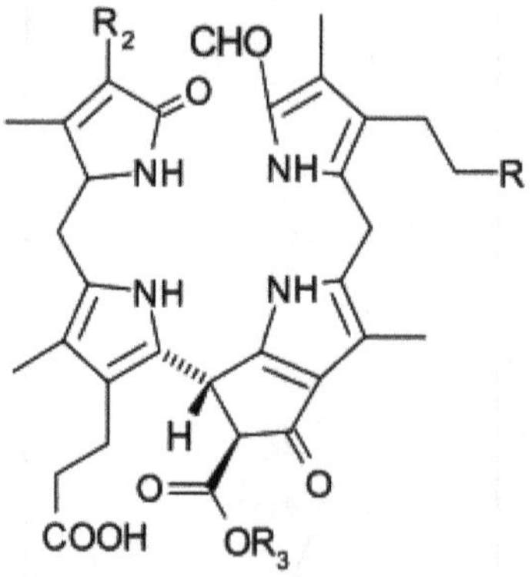

These compounds have also been identified in several ripening fruits.

Spectrophotometry

Measurement of the absorption of light is complicated by the solvent used to extract the chlorophyll from plant material, which affects the values obtained,

- In diethyl ether, chlorophyll a has approximate absorbance maxima of 430 nm and 662 nm, while chlorophyll b has approximate maxima of 453 nm and 642 nm.

- The absorption peaks of chlorophyll *a* are at 665 nm and 465 nm. Chlorophyll *a* fluoresces at 673 nm (maximum) and 726 nm. The peak molar absorption coefficient of chlorophyll *a* exceeds 10^5 M^{-1} cm^{-1}, which is among the highest for small-molecule organic compounds.

- In 90% acetone-water, the peak absorption wavelengths of chlorophyll *a* are 430 nm and 664 nm; peaks for chlorophyll *b* are 460 nm and 647 nm; peaks for chlorophyll *c1* are 442 nm and 630 nm; peaks for chlorophyll *c2* are 444 nm and 630 nm; peaks for chlorophyll *d* are 401 nm, 455 nm and 696 nm.

Because it absorbs red and blue light strongly but is transparent to green light, pure chlorophyll has a strong green colour.

By measuring the absorption of light in the red and far red regions it is possible to estimate the concentration of chlorophyll within a leaf.

In his scientific paper Gitelson (1999) states, "The ratio between chlorophyll fluorescence, at 735 nm and the wavelength range 700nm to 710 nm, F735/F700 was found to be linearly proportional to the chlorophyll content (with determination coefficient, r2, more than 0.95) and thus this ratio can be used as a precise indicator of chlorophyll content in plant leaves." The fluorescent ratio chlorophyll content meters use this technique.

Biosynthesis

In plants, chlorophyll may be synthesized from succinyl-CoA and glycine, although the immediate precursor to chlorophyll *a* and *b* is protochlorophyllide. In Angiosperm plants, the last step, conversion of protochlorophyllide to chlorophyll, is light-dependent and such plants are pale (etiolated) if grown in the darkness. Non-vascular plants and green algae have an additional light-independent enzyme and grow green in the darkness instead.

Chlorophyll itself is bound to proteins and can transfer the absorbed energy in the required direction. Protochlorophyllide occurs mostly in the free form and, under light conditions, acts as a photosensitizer, forming highly toxic free radicals. Hence, plants need an efficient mechanism of regulating the amount of chlorophyll precursor. In angiosperms, this is done at the step of aminolevulinic acid (ALA), one of the intermediate compounds in the biosynthesis pathway. Plants that are fed by ALA accumulate high and toxic levels of protochlorophyllide; so do the mutants with the damaged regulatory system.

Chlorosis is a condition in which leaves produce insufficient chlorophyll, turning them yellow. Chlorosis can be caused by a nutrient deficiency of iron—called iron chlorosis— or by a shortage of magnesium or nitrogen. Soil pH sometimes plays a role in nutrient-caused chlorosis; many plants are adapted to grow in soils with specific pH levels and their ability to absorb nutrients from the soil can be dependent on this. Chlorosis can also be caused by pathogens including viruses, bacteria and fungal infections, or sap-sucking insects.

Complementary Light Absorbance of Anthocyanins with Chlorophylls

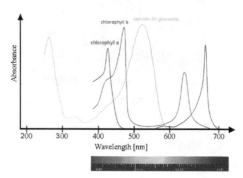

Superposition of spectra of chlorophyll a and b with oenin (malvidin 3O glucoside), a typical anthocyanidin, showing that, while chlorophylls absorb in the blue and yellow/red parts of the visible spectrum, oenin absorbs mainly in the green part of the spectrum, where chlorophylls don't absorb at all.

Anthocyanins are other plant pigments. The absorbance pattern responsible for the red color of anthocyanins may be complementary to that of green chlorophyll in photosynthetically active tissues such as young *Quercus coccifera* leaves. It may protect the leaves from attacks by plant eaters that may be attracted by green color.

Distribution

The chlorophyll maps show milligrams of chlorophyll per cubic meter of seawater each month. Places where chlorophyll amounts were very low, indicating very low numbers of phytoplankton, are blue. Places where chlorophyll concentrations were high, meaning many phytoplankton were growing, are yellow. The observations come from the Moderate Resolution Imaging Spectroradiometer (MODIS) on NASA's Aqua satellite. Land is dark gray, and places where MODIS could not collect data because of sea ice, polar darkness, or clouds are light gray.The highest chlorophyll concentrations, where tiny surface-dwelling ocean plants are thriving, are in cold polar waters or in places where ocean currents bring cold water to the surface, such as around the equator and along the shores of continents. It is not the cold water itself that stimulates the phytoplankton. Instead, the cool temperatures are often a sign that the water has welled up to the surface from deeper in the ocean, carrying nutrients that have built up over time. In polar waters, nutrients accumulate in surface waters during the dark winter months when plants cannot grow. When sunlight returns in the spring and summer, the plants flourish in high concentrations.

Culinary Use

Chlorophyll is registered as a food additive (colorant), and its E number is E140. Chefs use chlorophyll to color a variety of foods and beverages green, such as pasta and absinthe. Chlorophyll is not soluble in water, and it is first mixed with a small quantity of vegetable oil to obtain the desired solution.

Photopigment

Photopigments are unstable pigments that undergo a chemical change when they absorb light. The term is generally applied to the non-protein chromophore moiety of photosensitive chromoproteins, such as the pigments involved in photosynthesis and photoreception. In medical terminology, "photopigment" commonly refers to the photoreceptor proteins of the retina.

Photosynthetic Pigments

Photosynthetic pigment (converting light into biochemical energy). Examples for photosynthetic pigments are chlorophyll, carotenoids and phycobilins. These pigments enter a high-energy state upon absorbing a photon which they can release in the form of chemical energy. This can occur via light-driven pumping of ions across a biological membrane (e.g. in the case of the proton pump bacteriorhodopsin) or via excitation and transfer of electrons released by photolysis (e.g. in the photosystems of the thylakoid membranes of plant chloroplasts). In chloroplasts, the light-driven electron transfer chain in turn drives the pumping of protons across the membrane.

Photoreceptor Pigments

The pigments in photoreceptor proteins either change their conformation or undergo photoreduction when they absorb a photon. This change in the conformation or redox state of the chromophore then affects the protein conformation or activity and triggers a signal transduction cascade. Examples for photoreceptor pigments include retinal (for example in rhodopsin), flavin (for example in cryptochrome), and bilin (for example in phytochrome).

Photopigments of the Vertebrate Retina

In medical terminology, the term photopigment is applied to opsin-type photoreceptor proteins, specifically rhodopsin and photopsins, the photoreceptor proteins in the retinal rods and cones of vertebrates that are responsible for visual perception, but also melanopsin and others.

Phycobiliprotein

Phycobiliproteins are water-soluble proteins present in cyanobacteria and certain algae (rhodophytes, cryptomonads, glaucocystophytes) that capture light energy, which is then passed on to chlorophylls during photosynthesis. Phycobiliproteins are formed of a complex between proteins and covalently bound phycobilins that act as chromophores (the light-capturing part). They are most important constituents of the phycobilisomes.

Major phycobiliproteins:

Phyco-bilip-rotein	MW (kDa)	Ex (nm) / Em (nm)	Quan-tum yield	Molar Extinc-tion Co-efficient ($M^{-1}cm^{-1}$)	Comment
R-Phy-co-eryth-rin (R-PE)	240	498.546.566 nm / 576 nm	0,84	$1.53 \ 10^6$	Can be excited by Kr/Ar laser **Applications for R-Phycoerythrin** Many applications and instruments were developed specifically for R-phycoerythrin. It is commonly used in immunoassays such as FACS, flow cytometry, multimer/tetramer applications. **Structural Characteristics** R-phycoerythrin is also produced by certain red algae. The protein is made up of at least three different subunits and varies according to the species of algae that produces it. The subunit structure of the most common R-PE is $(\alpha\beta)_6\gamma$. The α subunit has two phycoerythrobilins (PEB), the β subunit has 2 or 3 PEBs and one phycourobilin (PUB), while the different gamma subunits are reported to have 3 PEB and 2 PUB (γ_1) or 1 or 2 PEB and 1 PUB (γ_2). (Phycobiliprotein overview information)

B-Phy-coeryth-rin (B-PE)	240	546.566 nm / 576 nm	0,98	(545 nm) 2.4 10^6 (563 nm) 2.33 10^6	**Applications for B-Phycoerythrin** Because of its high quantum yield, B-PE is considered the world's brightest fluorophore. It is compatible with commonly available lasers and gives exceptional results in flow cytometry, Luminex® and immunofluorescent staining. B-PE is also less "sticky" than common synthetic fluorophores and therefore gives less background interference. **Structural Characteristics** B-phycoerythrin (B-PE) is produced by certain red algae such as *Rhodella* sp. The specific spectral characteristics are a result of the composition of its subunits. B-PE is composed of at least three subunits and sometimes more. The chromophore distribution is as follows: α subunit with 2 phycoerythrobilins (PEB), β subunit with 3 PEB, and the γ subunit with 2 PEB and 2 phycourobilins (PUB). The quaternary structure is reported as $(\alpha\beta)_6\gamma$. (Phycobiliprotein overview information)
C-Phy-cocy-anin (CPC)	232	620 nm / 642 nm	0,81	1.54 10^6	Accepts the fluorescence for R-PE; Its red fluorescence can be transmitted to Allophycocyanin
Allo-phyco-cyanin (APC)	105	651 nm / 662 nm	0,68	7.3 10^5	Excited by He/Ne laser; double labeling with Sulfo-Rhodamine 101 or any other equivalent fluorochrome.Many applications and instruments were developed specifically for allophycocyanin. It is commonly used in immunoassays such as flow cytometry and high-throughput screening. It is also a common acceptor dye for FRET assays. **Structural Characteristics** Allophycocyanin can be isolated from various species of red or blue-green algae, each producing slightly different forms of the molecule. It is composed of two different subunits (α and β) in which each subunit has one phycocyanobilin (PCB) chromophore. The subunit structure for APC has been determined as $(\alpha\beta)_3$. (Phycobiliprotein overview information)
		↑ = FluoProbes PhycoBiliProteins data			

Characteristics and Applications in Biotechnology

Phycobiliproteins elicit great fluorescent properties compared to small organic fluoro-phores, especially when high sensitivity or multicolor detection is required :

- Broad and high absorption of light suits many light sources

- Very intense emission of light: 10-20 times brighter than small organic fluoro-phores

- Relative large Stokes shift gives low background, and allows multicolor detec-tions.

- Excitation and emission spectra do not overlap compared to conventional or-ganic dyes.

- Can be used in tandem (simultaneous use by FRET) with conventional chromo-phores (i.e. PE and FITC, or APC and SR101 with the same light source).

- Fluorescence retention period is longer.

- Very high water solubility

As a result, phycobiliproteins allow very high detection sensitivity, and can be used in various fluorescence based techniques fluorimetric microplate assays, Flow Cytometry, FISH and multicolor detection.

Bacteriochlorophyll

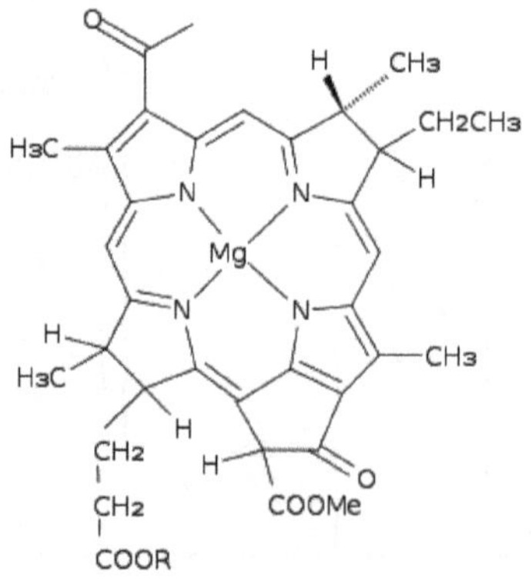

Bacteriochlorophyll *a*. R is phytyl or geranylgeranyl.

Bacteriochlorophylls are photosynthetic pigments that occur in various phototrophic bacteria. They were discovered by C. B. van Niel in 1932. They are related to chloro-phylls, which are the primary pigments in plants, algae, and cyanobacteria. Groups that contain bacteriochlorophyll conduct photosynthesis, but do not produce oxygen. They use wavelengths of light not absorbed by plants or Cyanobacteria. Different groups contain different types of bacteriochlorophyll:

Pigment	Bacterial group	*in vivo* infrared absorption maximum (nm)
Bacteriochlorophyll *a*	Purple bacteria, Heliobacteria, Green Sulfur Bacteria, Chloroflexi, *Chloracidobacterium thermophilum*	805, 830-890
Bacteriochlorophyll *b*	Purple bacteria	835-850, 1020-1040
Bacteriochlorophyll *c*	Green sulfur bacteria, Chloroflexi, *C. thermophilum*	745-755
Bacteriochlorophyll c_s	Chloroflexi	740
Bacteriochlorophyll *d*	Green sulfur bacteria	705-740
Bacteriochlorophyll *e*	Green sulfur bacteria	719-726
Bacteriochlorophyll *f*	Green sulfur bacteria (currently found only through mutation; natural may exist)	700-710
Bacteriochlorophyll *g*	Heliobacteria	670, 788

Bacteriochlorophylls *a*, *b*, and *g* are bacteriochlorins, meaning their molecules have a bacteriochlorin macrocycle ring with two reduced pyrrole rings (B and D). Bacteriochlorophylls *c*, *d*, *e*, and *f* are chlorins, meaning their molecules have a chlorin macrocycle ring with one reduced pyrrole ring (D).

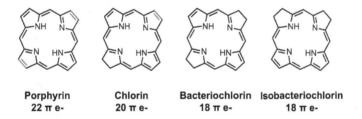

Porphyrin 22 π e-   Chlorin 20 π e-   Bacteriochlorin 18 π e-   Isobacteriochlorin 18 π e-

Chemical structures comparing porphin, chlorin, bacteriochlorin, and isobacteriochlorin. Note relocation of C=C double bond between the two bacteriochlorin isomers. There are two π electrons (symbolized by π e⁻) for every double bond in the macrocycle.

Phycoerythrin

Phycoerythrin (PE) is a red protein-pigment complex from the light-harvesting phycobiliprotein family, present in red algae and cryptophytes, accessory to the main chlorophyll pigments responsible for photosynthesis.

Like all phycobiliproteins, it is composed of a protein part covalently binding chromophores called phycobilins. In the phycoerythrin family, the most known phycobilins are: phycoerythrobilin, the typical phycoerythrin acceptor chromophore, and sometimes phycourobilin. Phycoerythrins are composed of $(\alpha\beta)$ monomers, usually organised in a disk-shaped trimer $(\alpha\beta)_3$ or hexamer $(\alpha\beta)_6$ (second one is the functional unit of the antenna rods). These typical complexes contain also third type of subunit, the γ chain.

The phycobiliproteins that bind the highest number of phycobilins (up to ten per $\alpha\beta$ monomer). (this non sequitur needs to be corrected

Phycobilisomes

Phycobiliproteins (like phycoerythrin) usually form rods of stacked disks in phycobilisomes.

Phycobiliproteins are part of huge light harvesting antennae protein complexes called phycobilisomes. In red algae they are anchored to the stromal side of thylakoid membranes of chloroplasts, whereas in cryptophytes phycobilisomes are reduced and (phycobiliprotein 545 PE545 molecules here) are densely packed inside the lumen of thylakoides.

Phycoerythrin is an accessory pigment to the main chlorophyll pigments responsible for photosynthesis. The light energy is captured by phycoerythrin and is then passed on to the reaction centre chlorophyll pair, most of the time via the phycobiliproteins phycocyanin and allophycocyanin.

Structural Characteristics

Phycoerythrins except phycoerythrin 545 (PE545) are composed of $(\alpha\beta)$ monomers assembled into disc-shaped $(\alpha\beta)_6$ hexamers or $(\alpha\beta)_3$ trimers with 32 or 3 symmetry and enclosing central channel. In phycobilisomes (PBS) each trimer or hexamer contains at least one linker protein located in central channel. B-phycoerythrin (B-PE) and R-phycoerythrin (R-PE) from red algae in addition to α and β chains have third, γ subunit combining linker and light-harvesting functions, because bears chromophores.

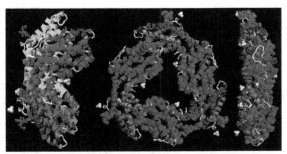

The crystal structure of B-phycoerythrin from red algae *Porphyridium cruentum* (PDB ID: 3V57). The asymmetric unit $(\alpha\beta)_2$ on the left and assumed biological molecule $(\alpha\beta)_3$. It contains phycoerythrobilin, N-methyl asparagine and SO_4^{2-}.

R-phycoerythrin is predominantly produced by red algae. The protein is made up of at least three different subunits and varies according to the species of algae that produces it. The subunit structure of the most common R-PE is $(\alpha\beta)_6\gamma$. The α subunit has two phycoerythrobilins (PEB), the β subunit has 2 or 3 PEBs and one phycourobilin (PUB), while the different gamma subunits are reported to have 3 PEB and 2 PUB (γ_1) or 1 or 2 PEB and 1 PUB (γ_2). The molecular weight of R-PE is 250,000 Daltons.

Crystal structures available in the Protein Data Bank contain in one $(\alpha\beta)_2$ or $(\alpha\beta\gamma)_2$ asymmetric unit of different phycoerythrins:

Phycoerythrobilin is the typical chromophore in phycoerythrin. It is similar to porphyrin of chlorophyll for example, but tetrapyrrole is linear, not closed into ring with metal ion in the middle.

The red algae *Gracilaria* contains R-phycoerythrin.

Chromophore or other non-protein molecule	Phycoerythrin				Chain
	PE545	B-PE	R-PE	other types	
Bilins	8	10	10	10	α and β

- Phycoerythrobilin (PEB)	6	10	0 or 8	8	β (PE545) or α and β
- 15,16-dihydrobiliverdin (DBV)	2	-	-	-	α (-3 and -2)
- Phycocyanobilin (CYC)	-	-	8 or 7 or 0	-	α and β
- Biliverdine IX alpha (BLA)	-	-	0 or 1	-	α
- Phycourobilin (PUB)	-	-	2	2	β
5-hydroxylysine (LYZ)	1 or 2	-		-	α (-3 or -3 and -2)
N-methyl asparagine (MEN)	2	2	0 or 2	2	β
Sulfate ion SO_4^{2-} (SO4)	-	5 or 1	0 or 2	-	α or α and β
Chloride ion Cl^- (CL)	1	-	-	-	β
Magnesium ion Mg^{2+} (MG)	2	-	-	-	α-3 and β
inspected PDB files	1XG0 1XF6 1QGW	3V57 3V58	1EYX 1LIA 1B8D	2VJH	

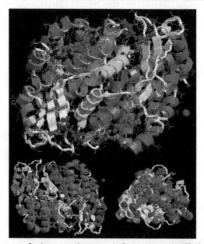

The crystal structure of phycoerythrin 545 (PE545) from a unicellular cryptophyte *Rhodomonas* CS24 (PDB ID: 1XG0). Colors: chains - alpha-2, alpha-3, beta, beta (helixes, sheets are yellow), phycoerythrobilin, 15,16-dihydrobiliverdin (15,16-DHBV), 5-hydroxylysine, N-methyl asparagine, Mg, Cl⁻.

The assumed biological molecule of phytoerythrin 545 (PE545) is $(αβ)_2$ or rather $(α_3β)$ $(α_2β)$. The numbers 2 and 3 after the α letters in second formula are part of chain names here, not their counts. The synonym cryptophytan name of $α_3$ chain is $α_1$ chain.

The largest assembly of B-phytoerythrin (B-PE) is $(αβ)_3$ trimer . However, preparations from red algae yield also $(αβ)_6$ hexamer . In case of R-phytoerythrin (R-PE) the largest assumed biological molecule here is $(αβγ)_6$, $(αβγ)_3(αβ)_3$ or $(αβ)_6$ dependently on publication, for other phytoeritrin types $(αβ)_6$. These γ chains from the Protein Data Bank are very small and consist only of three or six recognizable amino acids , whereas described at the beginning of this section linker γ chain is large (for example 277 amino acid long 33 kDa in case of $γ^{33}$ from red algae *Aglaothamnion neglectum*) This is because the electron density of the gamma-polypeptide is mostly averaged out by its threefold crystallographic symmetry and only a few amino acids can be modeled ·

For $(\alpha\beta\gamma)_6$, $(\alpha\beta)_6$ or $(\alpha\beta\gamma)_3(\alpha\beta)_3$ the values from the table should be simply multiplied by 3, $(\alpha\beta)_3$ contain intermediate numbers of non-protein molecules. (this non sequitur needs to be corrected)

In phycoerythrin PE545 above, one α chain (-2 or -3) binds one molecule of billin, in other examples it binds two molecules. The β chain always binds to three molecules. The small γ chain binds to none.

Two molecules of N-methyl asparagine are bound to the β chain, one 5-hydroxylysine to α (-3 or -2), one Mg^{2+} to α-3 and β, one Cl^- to β, 1-2 molecules of SO_4^{2-} to α or β.

Below is sample crystal structure of R-phycoerythrin from Protein Data Bank:

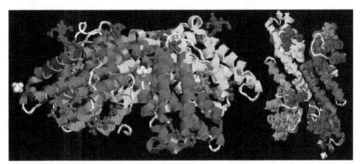

The crystal structure of R-phycoerythrin from red algae *Gracilaria chilensis* (PDB ID: 1EYX) - basic oligomer $(\alpha\beta\gamma)_2$ (so called asymmetric unit). It contains phycocyanobilin, biliverdine IX alpha, phycourobilin, N-methyl asparagine, SO_4^{2-}. One fragment of γ chain is red, second one white because it is not considered as alpha helix despite identical aminoacid sequence.

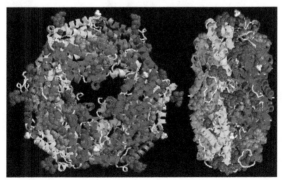

The entire oligomer of R-phycoerythrin from *Gracilaria chilensis* $(\alpha\beta\gamma)_6$ (PDB ID: 1EYX).

Spectral Characteristics

Absorption peaks in the visible light spectrum are measured at 495 and 545/566 nm, depending on the chromophores bound and the considered organism. A strong emission peak exists at 575 ± 10 nm. (*i.e.*, phycoerythrin absorbs slightly blue-green/yellowish light and emits slightly orange-yellow light.)

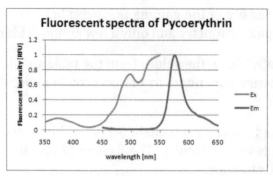

Fluorescent spectra of phycoerythrin

Property	Value
Absorption maximum	565 nm
Additional Absorption peak	498 nm
Emission maximum	573 nm
Extinction Coefficient (ε)	$1.96 \times 10^6 \text{ M}^{-1}\text{cm}^{-1}$
Quantum Yield (QY)	0.84
Brightness (ε x QY)	$1.65 \times 10^6 \text{ M}^{-1}\text{cm}^{-1}$

PEB and DBV bilins in PE545 absorb in the green spectral region too, with maxima at 545 and 569 nm respectively. The fluorescence emission maximum is at 580 nm.

R-phycoerythrin Variations

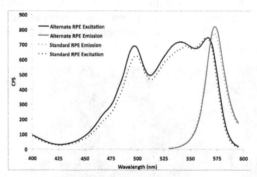

Excitation and emission profiles for r-phycoerythrin from two different algae. Common laser excitation wavelengths are also noted.

As mentioned above, phycoerythrin can be found in a variety of algal species. As such, there can be variation in the efficiency of absorbance and emission of light required for facilitation of photosynthesis. This could be a result of the depth in the water column that a specific alga typically resides and a consequent need for greater or less efficiency of the accessory pigments.

With advances in imaging and detection technology which can avoid rapid photo-bleaching, protein fluorophores have become a viable and powerful tool for researchers in fields such as microscopy, microarray analysis and Western blotting. In light of this, it may be beneficial for researchers to screen these variable R-phycoerythrins to determine which one is most appropriate for their particular application. Even a small increase in fluorescent efficiency could reduce background noise and lower the rate of false-negative results.

Practical Applications

R-Phycoerythrin, or PE, is useful in the laboratory as a fluorescence-based indicator for the presence of cyanobacteria and for labeling antibodies in a technique called immunofluorescence, among other applications. There are also other types of phyco-erythrins, such as B-Phycoerythrin, which have slightly different spectral properties. B-Phycoerythrin absorbs strongly at about 545 nm (slightly yellowish green) and emits strongly at 572 nm (yellow) instead and could be better suited for some instruments. B-Phycoerythrin may also be less "sticky" than R-Phycoerythrin and contributes less to background signal due to non-specific binding in certain applications.

R-Phycoerythrin and B-Phycoerythrin are among the brightest fluorescent dyes ever identified.

Phycocyanin

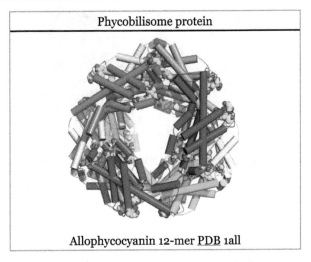

Phycobilisome protein

Allophycocyanin 12-mer PDB 1all

Phycocyanin is a pigment-protein complex from the light-harvesting phycobiliprotein family, along with allophycocyanin and phycoerythrin. It is an accessory pigment to chlorophyll. All phycobiliproteins are water-soluble, so they cannot exist within the membrane like carotenoids can. Instead, phycobiliproteins aggregate to form clusters that adhere to the membrane called phycobilisomes. Phycocyanin is a characteristic light blue color, absorbing orange and red light, particularly near 620 nm (depending on which specific type it is), and emits fluorescence at about 650 nm (also depending on which type it is). Allophycocyanin absorbs and emits at longer wavelengths than phycocyanin C or phycocyanin R. Phycocyanins are found in Cyanobacteria (also called blue-green algae). Phycobiliproteins have fluorescent properties that are used in immunoassay kits. Phycocyanin is from the Greek *phyco* meaning "algae" and *cyanin* is from the English word "cyan", which conventionally means a shade of blue-green (close to "aqua") and is derived from the Greek "kyanos" which means a somewhat different color: "dark blue". The product phycocyanin, produced by *Aphanizomenon flos-aquae* and spirulina, is for example used in the food and beverage industry as the natural coloring agent 'Lina Blue' and is found in sweets and ice cream.

The phycobiliproteins are made of two subunits(alpha and beta) having a protein backbone to which 1-2 linear tetrapyrrole chromophores are covalently bound.

C-phycocyanin is often found in cyanobacteria which thrive around hot springs, as it can be stable up to around 70°C, with identical spectroscopic (light absorbing) behaviours at 20 and 70°C. Thermophiles contain slightly different amino acid sequences making it stable under these higher conditions. Molecular weight is around 30,000 Da. Stability of this protein invitro at these temperatures has been shown to be substantially lower. Photo-spectral analysis of the protein after 1 min exposure to 65°C conditions in a purified state demonstrated a 50% loss of tertiary structure.

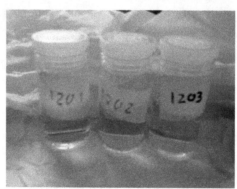

Phycocyanin pigment extracted from Microcystis Aeruginosa cyanobacteria.

Phycobilin

Phycobilins are light-capturing bilanes found in cyanobacteria and in the chloroplasts of red algae, glaucophytes and some cryptomonads (though not in green algae and plants). Most of their molecules consist of a chromophore which makes them colored. They are unique among the photosynthetic pigments in that they are bonded to certain water-soluble proteins, known as phycobiliproteins. Phycobiliproteins then pass the light energy to chlorophylls for photosynthesis.

The phycobilins are especially efficient at absorbing red, orange, yellow, and green light, wavelengths that are not well absorbed by chlorophyll *a*. Organisms growing in shallow waters tend to contain phycobilins that can capture yellow/red light, while those at greater depth often contain more of the phycobilins that can capture green light, which is relatively more abundant there.

The phycobilins fluoresce at a particular wavelength, and are, therefore, often used in research as chemical tags, e.g., by binding phycobiliproteins to antibodies in a technique known as immunofluorescence.

Types

There are four types of phycobilins:

1. Phycoerythrobilin, which is red

1. Phycourobilin, which is orange

2. Phycoviolobilin (also known as phycobiliviolin) found in phycoerythrocyanin

3. Phycocyanobilin (also known as phycobiliverdin), which is blue.

They can be found in different combinations attached to phycobiliproteins to confer specific spectroscopic properties.

Structural Relation to Other Molecules

In chemical terms, phycobilins consist of an open chain of four pyrrole rings (*tetrapyrrole*) and are structurally similar to the bile pigment bilirubin, which explains the name. (Bilirubin's conformation is also affected by light, a fact used for the phototherapy of jaundiced newborns.) Phycobilins are also closely related to the chromophores of the light-detecting plant pigment phytochrome, which also consist of an open chain of four pyrroles. Chlorophylls are composed of four pyrroles as well, but there the pyrroles are arranged in a ring and contain a metal atom in the center.

Allophycocyanin

Allophycocyanin is a protein from the light-harvesting phycobiliprotein family, along with phycocyanin, phycoerythrin and phycoerythrocyanin. It is an accessory pigment to chlorophyll. All phycobiliproteins are water-soluble and therefore cannot exist within the membrane like carotenoids, but aggregate forming clusters that adhere to the membrane called phycobilisomes. Allophycocyanin absorbs and emits red light (650 & 660 nm max, respectively), and is readily found in Cyanobacteria (also called blue-green algae), and red algae. Phycobilin pigments have fluorescent properties that are used in immunoassay kits. In flow cytometry, it is often abbreviated APC. To be effectively used in applications such as FACS, High-Throughput Screening (HTS) and microscopy, APC needs to be chemically cross-linked.

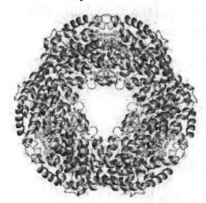

Allophycocyanin dodekamer, Gloeobacter violaceus.

Structural Characteristics

Allophycocyanin can be isolated from various species of red or blue-green algae, each producing slightly different forms of the molecule. It is composed of two different subunits (α and β) in which each subunit has one phycocyanobilin (PCB) chromophore. The subunit structure for APC has been determined as $(\alpha\beta)_3$. The molecular weight of APC is 105,000 Daltons.

Spectral Characteristics

Absorption maximum	652 nm
Additional Absorption peak	625 nm
Emission maximum	657.5 nm
Stokes Shift	5.5 nm
Extinction Coefficient	700,000 $M^{-1}cm^{-1}$
Quantum Yield	0.68
Brightness	1.6 x 10^5 $M^{-1}cm^{-1}$

Cross-linked APC

As mentioned above, in order for APC to be useful in immunoassays it must first be chemically cross-linked to prevent it from dissociating into its component subunits when in common physiological buffers. The conventional method for accomplishing this is through a destructive process wherein the treated APC trimer is chemically disrupted in 8M urea and then allowed to re-associate through in a physiological buffer. An alternative method can be used which preserves the structural integrity of the APC trimer and allows for a brighter, more stable end-product.

Applications

Many applications and instruments were developed specifically for Allophycocyanin. It is commonly used in immunoassays such as FACS, flow cytometry, and High Throughput Screening as an acceptor for Europium via Time Resolved-Fluorescence Resonance Energy Transfer (TR-FRET) assays.

References

- Pitchford, Paul (2002). Healing with Whole Foods: Asian Traditions and Modern Nutrition. North Atlantic Books. ISBN 1-55643-471-5.

- Simopoulos, Artemis P.; Gopalan, C., eds. (2003). Plants in Human Health and Nutrition Policy. Karger Publishers. ISBN 3-8055-7554-8.

- Gross, Jeana (1991). Pigments in vegetables: chlorophylls and carotenoids. Van Nostrand Reinhold, ISBN 0442006578.

- Larkum, edited by Anthony W. D. Larkum, Susan E. Douglas & John A. Raven (2003). Photosynthesis in algae. London: Kluwer. ISBN 0-7923-6333-7.

- Adams, Jad (2004). Hideous absinthe : a history of the devil in a bottle. United Kingdom: I.B.Tauris, 2004. p. 22. ISBN 1860649203.

- Australia New Zealand Food Standards Code"Standard 1.2.4 – Labelling of ingredients". Retrieved 2014-12-22.

- MicroPlate Detection comparison between SureLight®P-3L, other fluorophores and enzymatic detection Columbia Biosciences, 2010

Enzymes used in Photosynthesis

As the process of photosynthesis is highly complex, several specialized enzymes exist to facilitate it. This chapter deals with the enzymes such as alternative oxidase, plastid terminal oxidase and transketolase. Each of these enzymes play crucial roles in the various cycles of photosynthesis and the reader is provided in-depth information about how these enzymes interact with other biomolecules in cellular reactions.

Plastid Terminal Oxidase

Plastid terminal oxidase or plastoquinol terminal oxidase (PTOX) is an enzyme that resides on the thylakoid membranes of plant and algae chloroplasts and on the membranes of cyanobacteria. The enzyme was hypothesized to exist as a photosynthetic oxidase in 1982 and was verified by sequence similarity to the mitochondrial alternative oxidase (AOX).The two oxidases evolved from a common ancestral protein in prokaryotes, and they are so functionally and structurally similar that a thylakoid-localized AOX can restore the function of a PTOX knockout.

Function

Plastid terminal oxidase catalyzes the oxidation of the plastoquinone pool, which exerts a variety of effects on the development and functioning of plant chloroplasts.

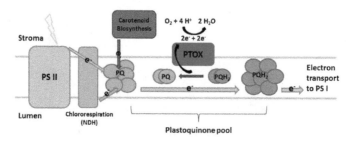

A summary of the pathways plastid terminal oxidase plays a role in through oxidation of the quinone pool

Carotenoid Biosynthesis and Plastid Development..

The enzyme is important for carotenoid biosynthesis during chloroplast biogenesis. In developing plastids, its activity prevents the over-reduction of the plastoquinone pool.

Knockout plants for PTOX exhibit phenotypes of variegated leaves with white patches. Without the enzyme, the carotenoid synthesis pathway slows down due to the lack of oxidized plastoquinone with which to oxidize phytoene, a carotenoid intermediate. The colorless compound phytoene accumulates in the leaves, resulting in white patches of cells.PTOX is also thought to determine the redox poise of the developing photosynthetic apparatus and without it, plants fail to assemble organized internal membrane structures in chloroplasts when exposed to high light during early development.

Photoprotection

Plants deficient in the *IMMUTANS* gene that encodes the oxidase are especially susceptible to photooxidative stress during early plastid development. The knockout plants exhibit a phenotype of variegated leaves with white patches that indicate a lack of pigmentation or photodamage. This effect is enhanced with increased light and temperature during plant development. The lack of plastid terminal oxidase indirectly causes photodamage during plastid development because protective carotenoids are not synthesized without the oxidase.

The enzyme is also thought to act as a safety valve for stress conditions in the photosynthetic apparatus. By providing an electron sink when the plastoquinone pool is over-reduced, the oxidase is thought to protect photosystem II from oxidative damage. Knockouts for Rubisco and photosystem II complexes, which would experience more photodamage than normal, exhibit an upregulation of plastid terminal oxidase.This effect is not universal because it requires plants to have additional PTOX regulation mechanisms. While many studies agree with the stress-protective role of the enzyme, one study showed that over-expression of *PTOX* increases the production of reactive oxygen species and causes more photodamage than normal. This finding suggests that an efficient antioxidant system is required for the oxidase to function as a safety valve for stress conditions and that it is more important during chloroplast biogenesis than in the regular functioning of the chloroplast.

Chlororespiration and Electron Flux

The most confirmed function of plastid terminal oxidase in developed chloroplasts is its role in chlororespiration. In this process, NADPH dehydrogenase (NDH) reduces the quinone pool and the terminal oxidase oxidizes it, serving the same function as cytochrome c oxidase from mitochondrial electron transport. In *Chlamydomonas*, there are two copies of the gene for the oxidase. *PTOX2* significantly contributes to the flux of electrons through chlororespiration in the dark.There is also evidence from experiments with tobacco that it functions in plant chlororespiration as well.

In fully developed chloroplasts, prolonged exposure to light increases the activity of the oxidase. Because the enzyme acts at the plastoquinone pool in between photosystem II and photosystem I, it may play a role in controlling electron flow through photosynthesis by acting as an alternative electron sink. Similar to its role in carotenoid synthesis, its oxidase activity may prevent the over-reduction of photosystem I electron acceptors and damage by photoinhibition. A recent analysis of electron flux through the photosynthetic pathway shows that even when activated, the electron flux plastid terminal oxidase diverts is two orders of magnitude less than the total flux through photosynthetic electron transport. This suggests that the protein may play less of a role than previously thought in relieving the oxidative stress in photosynthesis.

Structure

Plastid terminal oxidase is an integral membrane protein, or more specifically, an integral monotopic protein and is bound to the thylakoid membrane facing the stroma. Based on sequence homology, the enzyme is predicted to contain four alpha helix domains that encapsulate a di-iron center. The two iron atoms are ligated by six essential conserved histidine and glutamate residues – Glu136, Glu175, His171, Glu227, Glu296, and His299. The predicted structure is similar to that of the alternative oxidase, with an additional Exon 8 domain that is required for the plastid oxidase's activity and stability. The enzyme is anchored to the membrane by a short fifth alpha helix that contains a Tyr212 residue hypothesized to be involved in substrate binding.

Mechanism

The oxidase catalyzes the transfer of four electrons from reduced plastoquinone to molecular oxygen to form water . The net reaction is written below:

$$2\ QH_2 + O_2 \rightarrow 2\ Q + 2\ H_2O$$

Analysis of substrate specificity revealed that the enzyme almost exclusively catalyzes the reduction of plastoquinone over other quinones such as ubiquinone and duroquinone. Additionally, iron is essential for the catalytic function of the enzyme and cannot be substituted by another metal cation like Cu^{2+}, Zn^{2+}, or Mn^{2+} at the catalytic center.@It is unlikely that four electrons could be transferred at once in a single iron cluster, so all of the proposed mechanisms involve two separate two-electron transfers from reduced plastoquinone to the di-iron center. In the first step common to all proposed mechanisms, one plastoquinone is oxidized and both irons are reduced from iron(III) to iron(II). Four different mechanisms are proposed for the next step, oxygen capture. One mechanism proposes a peroxide intermediate, after which one oxygen atom is used to create water and another is left bound in a diferryl configuration. Upon one more plastoquinone oxidation, a second water molecule is formed and the irons return to a +3 oxidation state. The other mechanisms involve the formation of Fe(III)-OH or Fe(IV)-OH and a tyrosine radical. These radical-based mechanisms could explain why over-expression of the *PTOX* gene causes increased generation of reactive oxygen species.

Evolution

The enzyme is present in organisms capable of oxygenic photosynthesis, which includes plants, algae, and cyanobacteria. Plastid terminal oxidase and alternative oxidase are thought to have originated from a common ancestral di-iron carboxylate protein. Oxygen reductase activity was likely an ancient mechanism to scavenge oxygen in the early transition from an anaerobic to aerobic world. The plastid oxidase first evolved in ancient cyanobacteria and the alternative oxidase in proteobacteria before eukaryotic evolution and endosymbiosis events. Through endosymbiosis, the plastid oxidase was vertically inherited by eukaryotes that evolved into plants and algae. Sequenced genomes of various plant and algae species shows that the amino acid sequence is more than 25% conserved, which is a significant amount of conservation for an oxidase. This sequence conservation further supports the theory that both the alternative and plastid oxidases evolved before endosymbiosis and did not significantly change through eukaryote evolution.

There also exist PTOX cyanophages that contain copies of the gene for the plastid oxidase. They are known to act as viral vectors for movement of the gene between cyanobacterial species. Some evidence suggests that the phages may use the oxidase to influence photosynthetic electron flow to produce more ATP and less NADPH because viral synthesis utilizes more ATP.

Alternative Oxidase

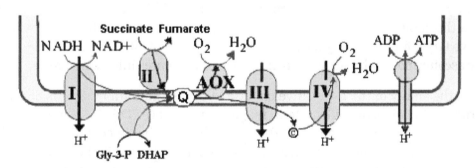

The alternative oxidase shown as part of the complete electron transport chain. UQ is ubiquinol/ubiquinone, C is cytochrome c and AOX is the alternative oxidase.

The alternative oxidase (AOX) is an enzyme that forms part of the electron transport chain in mitochondria of different organisms Proteins homologous to the mitochondrial oxidase have also been identified in bacterial genomes.

The oxidase provides an alternative route for electrons passing through the electron transport chain to reduce oxygen. However, as several proton-pumping steps are bypassed in this alternative pathway, activation of the oxidase reduces ATP generation. This enzyme was first identified as a distinct oxidase pathway from cytochrome c oxidase as the alternative oxidase is resistant to inhibition by the poison cyanide.

Function

The fungicide azoxystrobin.

This metabolic pathway leading to the alternative oxidase diverges from the cytochrome-linked electron transport chain at the ubiquinone pool. Alternative pathway respiration only produces proton translocation at Complex 1 (NADH dehydrogenase) and so has a lower ATP yield than the full pathway. The expression of the alternative oxidase gene *AOX* is influenced by stresses such as cold, reactive oxygen species and infection by pathogens, as well as other factors that reduce electron flow through the cytochrome pathway of respiration. Although the benefit conferred by this activity remains uncertain, it may enhance an organisms' ability to resist these stresses, through reducing the level of oxidative stress.

Unusually, the bloodstream form of the protozoan parasite *Trypanosoma brucei*, which is the cause of sleeping sickness, depends entirely on the alternative oxidase pathway for cellular respiration through its electron transport chain. This major metabolic difference between the parasite and its human host has made the *T. brucei* alternative oxidase an attractive target for drug design. Of the known inhibitors of alternative oxidases, the antibiotic ascofuranone inhibits the *T. brucei* enzyme and cures infection in mice.

In fungi, the ability of the alternative oxidase to bypass inhibition of parts of the electron transport chain can contribute to fungicide resistance. This is seen in the strobilurin fungicides that target complex III, such as azoxystrobin, picoxystrobin and fluoxastrobin. However, as the alternative pathway generates less ATP, these fungicides are still effective in preventing spore germination, as this is an energy-intensive process.

Structure and Mechanism

The alternative oxidase is an integral monotopic membrane protein that is tightly bound to the inner mitochondrial membrane from matrix side The enzyme has been predicted to contain a coupled diiron center on the basis of a conserved sequence motif consisting of the proposed iron ligands, four glutamate and two histidine amino acid residues. The electron spin resonance study of *Arabidopsis thaliana* alternative oxidase AOX1a showed that the enzyme contains a hydroxo-bridged mixed-valent Fe(II)/Fe(III) binuclear iron center. A catalytic cycle has been proposed that involves this di-iron center and at least one transient protein-derived free radical, which is probably formed on a tyrosine residue.

Transketolase

transketolase

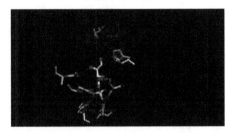

Transketolase is an enzyme of both the pentose phosphate pathway in all organisms and the Calvin cycle of photosynthesis. It catalyzes two important reactions, which operate in opposite directions in these two pathways. In the first reaction of the non-oxidative pentose phosphate pathway, the cofactor thiamine diphosphate accepts a 2-carbon fragment from a 5-carbon ketose (D-xylulose-5-P), then transfers this fragment to a 5-carbon aldose (D-ribose-5-P) to form a 7-carbon ketose (sedoheptulose-7-P). The abstraction of two carbons from D-xylulose-5-P yields the 3-carbon aldose glyceraldehyde-3-P. In the Calvin cycle, transketolase catalyzes the reverse reaction, the conversion of sedoheptulose-7-P and glyceraldehyde-3-P to pentoses, the aldose D-ribose-5-P and the ketose D-xylulose-5-P.

The second reaction catalyzed by transketolase in the pentose phosphate pathway involves the same thiamine diphosphate-mediated transfer of a 2-carbon fragment from D-xylulose-5-P to the aldose erythrose-4-phosphate, affording fructose 6-phosphate and glyceraldehyde-3-P. Again, in the Calvin cycle exactly the same reaction occurs, but in the opposite direction. Moreover, in the Calvin cycle this is the first reaction catalyzed by transketolase, rather than the second.

In mammals, transketolase connects the pentose phosphate pathway to glycolysis, feeding excess sugar phosphates into the main carbohydrate metabolic pathways. Its presence is necessary for the production of NADPH, especially in tissues actively engaged in biosyntheses, such as fatty acid synthesis by the liver and mammary glands, and for steroid synthesis by the liver and adrenal glands. Thiamine diphosphate is an essential cofactor, along with calcium.

Transketolase is abundantly expressed in the mammalian cornea by the stromal keratocytes and epithelial cells and is reputed to be one of the corneal crystallins.

Species Distribution

Transketolase is widely expressed in a wide range of organisms including bacteria, plants, and mammals. The following human genes encode proteins with transketolase activity:

- *TKT* (transketolase)

- *TKTL1* (transketolase-like protein 1)

- *TKTL2* (transketolase-like protein 2)

Structure

The entrance to the active site for this enzyme is made up mainly of several arginine, histidine, serine, and aspartate side-chains, with a glutamate side-chain playing a secondary role. These side-chains, to be specific Arg359, Arg528, His469, and Ser386, are conserved within each transketolase enzyme and interact with the phosphate group of the donor and acceptor substrates. Because the substrate channel is so narrow, the donor and acceptor substrates cannot be bound simultaneously. Also, the substrates conform into a slightly extended form upon binding in the active site to accommodate this narrow channel.

Although this enzyme is able to bind numerous types of substrates, such as phosphorylated and nonphosphorylated monosaccharides including the keto and aldosugars fructose, ribose, etc., it has a high specificity for the stereoconfiguration of the hydroxyl groups of the sugars. These hydroxyl groups at C-3 and C-4 of the ketose donor must be in the D-*threo* configuration in order to correctly correspond to the C-1 and C-2 positions on the aldose acceptor. Also they stabilize the substrate in the active site by interacting with the Asp477, His30, and His263 residues. Disruption of this configuration, both the placement of hydroxyl groups or their stereochemistry, would consequently alter the H-bonding between the residues and substrates thus causing a lower affinity for the substrates.

In the first half of this pathway, His263 is used to effectively abstract the C3 hydroxyl proton, which thus allows a 2-carbon segment to be cleaved from fructose 6-phosphate. The cofactor necessary for this step to occur is thiamin pyrophosphate (TPP). The binding of TPP to the enzyme incurs no major conformational change to the enzyme; instead, the enzyme has two flexible loops at the active site that make TPP accessible and binding possible. Thus, this allows the active site to have a "closed" conformation rather than a large conformational change. Later in the pathway, His263 is used as a proton donor for the substrate acceptor-TPP complex, which can then generate erythrose-4-phosphate.

The histidine and aspartate side-chains are used to effectively stabilize the substrate within the active site and also participate in deprotonation of the substrate. To be specific, the His 263 and His30 side-chains form hydrogen bonds to the aldehyde end of the substrate, which is deepest into the substrate channel, and Asp477 forms hydrogen bonds with the alpha hydroxyl group on the substrate, where it works to effectively bind the substrate and check for proper stereochemistry. It is also thought that Asp477 could have important catalytic effects because of its orientation in the middle of the active site and its interactions with the alpha hydroxyl group of the substrate. Glu418, which is located in the deepest region of the active site, plays a critical role in stabilizing the TPP cofactor. To be specific, it is involved in the cofactor-assisted proton abstraction from the substrate molecule.

The phosphate group of the substrate also plays an important role in stabilizing the substrate upon its entrance into the active site. The tight ionic and polar interactions between this phosphate group and the residues Arg359, Arg528, His469, and Ser386 collectively work to stabilize the substrate by forming H-bonds to the oxygen atoms of the phosphate. The ionic nature is found in the salt bridge formed from Arg359 to the phosphate group.

Mechanism

The catalysis of this mechanism is initiated by the deprotonation of TPP at the thiazolium ring. This carbanion then binds to the carbonyl of the donor substrate thus cleaving the bond between C-2 and C-3. This keto fragment remains covalently bound to the C-2 carbon of TPP. The donor substrate is then released, and the acceptor substrate enters the active site where the fragment, which is bound to the intermediate α-β-dihydroxyethyl thiamin diphosphate, is then transferred to the acceptor.

Experiments have also been conducted that test the effect replacing alanine for the amino acids at the entrance to the active site, Arg359, Arg528, and His469, which interact with the phosphate group of the substrate. This replacement creates a mutant enzyme with impaired catalytic activity.

Role in Disease

Transketolase activity is decreased in deficiency of thiamine, which in general is due to malnutrition. Several diseases are associated with thiamine deficiency, including beriberi, Biotin-Thiamine-Responsive Basal Ganglia Disease, Wernicke-Korsakoff syndrome, and others.

In Wernicke-Korsakoff syndrome, while no mutations could be demonstrated, there is an indication that thiamine deficiency leads to Wernicke-Korsakoff syndrome only in those whose transketolase has a reduced affinity for thiamine. In this way, the activity of transketolase is greatly hindered, and, as a consequence, the entire pentose phosphate pathway is inhibited.

Diagnostic Use

Red blood cell transketolase activity is reduced in deficiency of thiamine (vitamin B_1), and may be used in the diagnosis of Wernicke's encephalopathy and other B_1-deficiency syndromes if the diagnosis is in doubt. Apart from the baseline enzyme activity (which may be normal even in deficiency states), acceleration of enzyme activity after the addition of thiamine pyrophosphate may be diagnostic of thiamine deficiency (0-15% normal, 15-25% deficiency, >25% severe deficiency).

References

- Cox, Michael; Nelson, David R.; Lehninger, Albert L (2005). Lehninger principles of biochemistry. San Francisco: W.H. Freeman. ISBN 0-7167-4339-6.

- Sun X, Wen T (December 2011). "Physiological roles of plastid terminal oxidase in plant stress responses". J. Biosci. 36 (5): 951–6. doi:10.1007/s12038-011-9161-7. PMID 22116293.

- Houille-Vernes L, Rappaport F, Wollmann FA, Alric J, Johnson X (December 2011). "Plastid terminal oxidase 2 (PTOX2) is the major oxidase involved in chlororespiration in Chlamydomonas". PNAS. 106 (51): 20820–20825. doi:10.1073/pnas.1110518109.

Photosynthetic Efficiency and Chemical Reactions

When chlorophyll absorbs light, it sets into motion a series of reduction-oxidation reactions that ultimately result in the production of energy in the form of ATP molecules. Photosynthetic efficiency refers to the fraction of light that is converted into energy. This chapter provides information on light-dependent reactions, light-independent reactions, C4 carbon fixation, photodissociation and oxygen evolution. The reader is able to better understand the relevance each of these reactions has on photosynthesis.

Photosynthetic Efficiency

The photosynthetic efficiency is the fraction of light energy converted into chemical energy during photosynthesis in plants and algae. Photosynthesis can be described by the simplified chemical reaction

$$6H_2O + 6CO_2 + \text{energy} \rightarrow C_6H_{12}O_6 + 6O_2$$

where $C_6H_{12}O_6$ is glucose (which is subsequently transformed into other sugars, cellulose, lignin, and so forth). The value of the photosynthetic efficiency is dependent on how light energy is defined – it depends on whether we count only the light that is absorbed, and on what kind of light is used. It takes eight (or perhaps 10 or more) photons to utilize one molecule of CO_2. The Gibbs free energy for converting a mole of CO_2 to glucose is 114 kcal, whereas eight moles of photons of wavelength 600 nm contains 381 kcal, giving a nominal efficiency of 30%. However, photosynthesis can occur with light up to wavelength 720 nm so long as there is also light at wavelengths below 680 nm to keep Photosystem II operating. Using longer wavelengths means less light energy is needed for the same number of photons and therefore for the same amount of photosynthesis. For actual sunlight, where only 45% of the light is in the photosynthetically active wavelength range, the theoretical maximum efficiency of solar energy conversion is approximately 11%. In actuality, however, plants do not absorb all incoming sunlight (due to reflection, respiration requirements of photosynthesis and the need for optimal solar radiation levels) and do not convert all harvested energy into biomass, which results in an overall photosynthetic efficiency of 3 to 6% of total solar radiation. If photosynthesis is inefficient, excess light energy must be dissipated to avoid damaging the photosynthetic apparatus. Energy can be dissipated as heat (non-photochemical quenching), or emitted as chlorophyll fluorescence.

Typical Efficiencies

Plants

Quoted values sunlight-to-biomass efficiency

Plant	Efficiency
Plants, typical	0.1% 0.2–2%
Typical crop plants	1–2%
Sugarcane	7–8% peak

The following is a breakdown of the energetics of the photosynthesis process from *Photosynthesis* by Hall and Rao:

Starting with the solar spectrum falling on a leaf,

47% lost due to photons outside the 400–700 nm active range (chlorophyll utilizes photons between 400 and 700 nm, extracting the energy of one 700 nm photon from each one)

30% of the in-band photons are lost due to incomplete absorption or photons hitting components other than chloroplasts

24% of the absorbed photon energy is lost due to degrading short wavelength photons to the 700 nm energy level

68% of the utilized energy is lost in conversion into d-glucose

35–45% of the glucose is consumed by the leaf in the processes of dark and photo respiration

Stated another way:

100% sunlight → non-bioavailable photons waste is 47%, leaving

53% (in the 400–700 nm range) → 30% of photons are lost due to incomplete absorption, leaving

37% (absorbed photon energy) → 24% is lost due to wavelength-mismatch degradation to 700 nm energy, leaving

28.2% (sunlight energy collected by chlorophyl) → 32% efficient conversion of ATP and NADPH to d-glucose, leaving

9% (collected as sugar) → 35–40% of sugar is recycled/consumed by the leaf in dark and photo-respiration, leaving

5.4% net leaf efficiency.

Many plants lose much of the remaining energy on growing roots. Most crop plants store ~0.25% to 0.5% of the sunlight in the product (corn kernels, potato starch, etc.). Sugar cane is exceptional in several ways, yielding peak storage efficiencies of ~8%.

Measuring the photosynthetic efficiency of wheat in the field using an LCpro-SD

Photosynthesis increases linearly with light intensity at low intensity, but at higher intensity this is no longer the case. Above about 10,000 lux or ~100 watts/square meter the rate no longer increases. Thus, most plants can only utilize ~10% of full mid-day sunlight intensity. This dramatically reduces average achieved photosynthetic efficiency in fields compared to peak laboratory results. However, real plants (as opposed to laboratory test samples) have lots of redundant, randomly oriented leaves. This helps to keep the average illumination of each leaf well below the mid-day peak enabling the plant to achieve a result closer to the expected laboratory test results using limited illumination.

Only if the light intensity is above a plant specific value, called the compensation point the plant assimilates more carbon and releases more oxygen by photosynthesis than it consumes by cellular respiration for its own current energy demand. Photosynthesis measurement systems are not designed to directly measure the amount of light absorbed by the leaf. Nevertheless, the light response curves that the class produces do allow comparisons in photosynthetic efficiency between plants.

Algae and Other Monocellular Organisms

From a 2010 study by the University of Maryland, photosynthesizing Cyanobacteria have been shown to be a significant species in the global carbon cycle, accounting for 20–30% of Earth's photosynthetic productivity and convert solar energy into biomass-stored chemical energy at the rate of ~450 TW.

Worldwide Figures

According to the cyanobacteria study above, this means the total photosynthetic productivity of earth is between ~1500–2250 TW, or 47,300–71,000 exajoules per year. Using this source's figure of 178,000 TW of solar energy hitting the Earth's surface, the total photosynthetic efficiency of the planet is 0.84% to 1.26% .

Efficiencies of Various Biofuel Crops

Popular choices for plant biofuels include: oil palm, soybean, castor oil, sunflower oil, safflower oil, corn ethanol, and sugar cane ethanol.

An analysis of a proposed Hawaiian oil palm plantation claimed to yield 600 gallons of biodiesel per acre per year. That comes to 2835 watts per acre or 0.7 W/m^2. Typical insolation in Hawaii is around 5.5 $kWh/(m^2 day)$ or 230 W/m^2. For this particular oil palm plantation, if it delivered the claimed 600 gallons of biodiesel per acre per year, would be converting 0.3% of the incident solar energy to chemical fuel. Total photosynthetic efficiency would include more than just the biodiesel oil, so this 0.3% number is something of a lower bound.

Contrast this with a typical photovoltaic installation, which would produce an average of roughly 22 W/m^2 (roughly 10% of the average insolation), throughout the year. Furthermore, the photovoltaic panels would produce electricity, which is a high-quality form of energy, whereas converting the biodiesel into mechanical energy entails the loss of a large portion of the energy. On the other hand, a liquid fuel is much more convenient for a vehicle than electricity, which has to be stored in heavy, expensive batteries.

Most crop plants store ~0.25% to 0.5% of the sunlight in the product (corn kernels, potato starch, etc.), sugar cane is exceptional in several ways to yield peak storage efficiencies of ~8%.

Ethanol fuel in Brazil has a calculation that results in: "Per hectare per year, the biomass produced corresponds to 0.27 TJ. This is equivalent to 0.86 W/m^2. Assuming an average insolation of 225 W/m^2, the photosynthetic efficiency of sugar cane is 0.38%." Sucrose accounts for little more than 30% of the chemical energy stored in the mature plant; 35% is in the leaves and stem tips, which are left in the fields during harvest, and 35% are in the fibrous material (bagasse) left over from pressing.

C3 vs. C4 and CAM Plants

C3 plants use the Calvin cycle to fix carbon. C4 plants use a modified Calvin cycle in which they separate Ribulose-1,5-bisphosphate carboxylase oxygenase (RuBisCO) from atmospheric oxygen, fixing carbon in their mesophyll cells and using oxaloacetate and malate to ferry the fixed carbon to RuBisCO and the rest of the Calvin cycle enzymes isolated in the bundle-sheath cells. The intermediate compounds both contain four carbon atoms, which gives C4. In Crassulacean acid metabolism (CAM), time isolates functioning RuBisCo (and the other Calvin cycle enzymes) from high oxygen concentrations produced by photosynthesis, in that O_2 is evolved during the day, and allowed to dissipate then, while at night atmospheric CO_2 is taken up and stored as malic or other acids. During the day, CAM plants close stomata and use stored acids as carbon sources for sugar, etc. production.

The C3 pathway requires 18 ATP for the synthesis of one molecule of glucose while the C4 pathway requires 20 ATP. Despite this reduced ATP efficiency, C4 is an evolutionary advancement, adapted to areas of high levels of light, where the reduced ATP efficiency is more than offset by the use of increased light. The ability to thrive despite restricted water availability maximizes the ability to use available light. The simpler C3 cycle which operates in most plants is adapted to wetter darker environments, such as many northern latitudes.. Corn, sugar cane, and sorghum are C4 plants. These plants are economically important in part because of their relatively high photosynthetic efficiencies compared to many other crops. Pineapple is a CAM plant.

Light-dependent Reactions

In photosynthesis, the light-dependent reactions take place on the thylakoid membranes. The inside of the thylakoid membrane is called the lumen, and outside the thylakoid membrane is the stroma, where the light-independent reactions take place. The thylakoid membrane contains some integral membrane protein complexes that catalyze the light reactions. There are four major protein complexes in the thylakoid membrane: Photosystem II (PSII), Cytochrome b6f complex, Photosystem I (PSI), and ATP synthase. These four complexes work together to ultimately create the products ATP and NADPH.

The two photosystems absorb light energy through pigments - primarily the chlorophylls, which are responsible for the green color of leaves. The light-dependent reactions begin in photosystem II. When a chlorophyll a molecule within the reaction center of PSII absorbs a photon, an electron in this molecule attains a higher energy level. Because this state of an electron is very unstable, the electron is transferred from one to another molecule creating a chain of redox reactions, called an electron transport chain (ETC). The electron flow goes from PSII to cytochrome b6f to PSI. In PSI, the electron gets the energy from another photon. The final electron acceptor is NADP. In oxygenic photosynthesis, the first electron donor is water, creating oxygen as a waste product. In anoxygenic photosynthesis various electron donors are used.

Cytochrome b6f and ATP synthase work together to create ATP. This process is called photophosphorylation, which occurs in two different ways. In non-cyclic photophosphorylation, cytochrome b6f uses the energy of electrons from PSII to pump protons from the stroma to the lumen. The proton gradient across the thylakoid membrane creates a proton-motive force, used by ATP synthase to form ATP. In cyclic photophosphorylation, cytochrome b6f uses the energy of electrons from not only PSII but also PSI to create more ATP and to stop the production of NADPH. Cyclic phosphorylation is important to create ATP and maintain NADPH in the right proportion for the light-independent reactions.

The net-reaction of all light-dependent reactions in oxygenic photosynthesis is:

$$2H_2O + 2NADP^+ + 3ADP + 3P_i \rightarrow O_2 + 2NADPH + 3ATP$$

The two photosystems are protein complexes that absorb photons and are able to use this energy to create an electron transport chain. Photosystem I and II are very similar in structure and function. They use special proteins, called light-harvesting complexes, to absorb the photons with very high effectiveness. If a special pigment molecule in a photosynthetic reaction center absorbs a photon, an electron in this pigment attains the excited state and then is transferred to another molecule in the reaction center. This reaction, called photoinduced charge separation, is the start of the electron flow and is unique because it transforms light energy into chemical forms.

The Reaction Center

The reaction center is in the thylakoid membrane. It transfers light energy to a dimer of chlorophyll pigment molecules near the periplasmic (or thylakoid lumen) side of the membrane. This dimer is called a special pair because of its fundamental role in photosynthesis. This special pair is slightly different in PSI and PSII reaction center. In PSII, it absorbs photons with a wavelength of 680 nm, and it is therefore called P680. In PSI, it absorbs photons at 700 nm, and it is called P700. In bacteria, the special pair is called P760, P840, P870, or P960.

If an electron of the special pair in the reaction center becomes excited, it cannot transfer this energy to another pigment using resonance energy transfer. In normal circumstances, the electron should return to the ground state, but, because the reaction center is arranged so that a suitable electron acceptor is nearby, the excited electron can move from the initial molecule to the acceptor. This process results in the formation of a positive charge on the special pair (due to the loss of an electron) and a negative charge on the acceptor and is, hence, referred to as photoinduced charge separation. In other words, electrons in pigment molecules can exist at specific energy levels. Under normal circumstances, they exist at the lowest possible energy level they can. However, if there is enough energy to move them into the next energy level, they can absorb that energy and occupy that higher energy level. The light they absorb contains the necessary amount of energy needed to push them into the next level. Any light that does not have enough or has too much energy cannot be absorbed and is reflected. The electron in the higher energy level, however, does not want to be there; the electron is unstable and must return to its normal lower energy level. To do this, it must release the energy that has put it into the higher energy state to begin with. This can happen various ways. The extra energy can be converted into molecular motion and lost as heat. Some of the extra energy can be lost as heat energy, while the rest is lost as light. This re-emission of light energy is called fluorescence. The energy, but not the e- itself, can be passed onto another molecule. This is called resonance. The energy and the e- can be transferred to another molecule. Plant pigments usually utilize the last two of these reactions to convert the sun's energy into their own.

This initial charge separation occurs in less than 10 picoseconds (10^{-11} seconds). In their high-energy states, the special pigment and the acceptor could undergo charge recombination; that is, the electron on the acceptor could move back to neutralize the positive charge on the special pair. Its return to the special pair would waste a valuable high-energy electron and simply convert the absorbed light energy into heat. In the case of PSII, this backflow of electrons can produce reactive oxygen species leading to photoinhibition. Three factors in the structure of the reaction center work together to suppress charge recombination nearly completely.

- Another electron acceptor is less than 10 Å away from the first acceptor, and so the electron is rapidly transferred farther away from the special pair.

- An electron donor is less than 10 Å away from the special pair, and so the positive charge is neutralized by the transfer of another electron

- The electron transfer back from the electron acceptor to the positively charged special pair is especially slow. The rate of an of electron transfer reaction increases with its thermodynamic favorability up to a point and then decreases. The back transfer is so favourable that it takes place in the inverted region where electron-transfer rates become slower.

Thus, electron transfer proceeds efficiently from the first electron acceptor to the next, creating an electron transport chain that ends if it has reached NADPH.

Photosynthetic Electron Transport Chains in Chloroplasts

The photosynthesis process in chloroplasts begins when an electron of P680 of PSII attains a higher-energy level. This energy is used to reduce a chain of electron acceptors that have subsequently lowered redox-potentials. This chain of electron acceptors is known as an electron transport chain. When this chain reaches PS I, an electron is again excited, creating a high redox-potential. The electron transport chain of photosynthesis is often put in a diagram called the z-scheme, because the redox diagram from P680 to P700 resembles the letter z.

The final product of PSII is plastoquinol, a mobile electron carrier in the membrane. Plastoquinol transfers the electron from PSII to the proton pump, cytochrome b6f. The ultimate electron donor of PSII is water. Cytochrome b6f proceeds the electron chain to PSI through plastocyanin molecules. PSI is able to continue the electron transfer in two different ways. It can transfer the electrons either to plastoquinol again, creating a cyclic electron flow, or to an enzyme called FNR (Ferredoxin—NADP(+) reductase), creating a non-cyclic electron flow. PSI releases FNR into the stroma, where it reduces NADP+ to NADPH.

Activities of the electron transport chain, especially from cytochrome b6f, lead to pumping of protons from the stroma to the lumen. The resulting transmembrane proton gradient is used to make ATP via ATP synthase.

The overall process of the photosynthetic electron transport chain in chloroplasts is:

$$H_2O \rightarrow PS\ II \rightarrow plastoquinone \rightarrow cyt^b_6 \rightarrow plastocyanin \rightarrow PS\ I \rightarrow NADPH$$

Photosystem II

PS II is an extremely complex, highly organized transmembrane structure that contains a *water-splitting complex*, chlorophylls and carotenoid pigments, a *reaction center* (P680), pheophytin (a pigment similar to chlorophyll), and two quinones. It uses the energy of sunlight to transfer electrons from water to a mobile electron carrier in the membrane called *plastoquinone*:

$$H_2O \rightarrow P680 \rightarrow P680^* \rightarrow plastoquinone$$

Plastoquinone, in turn, transfers electrons to *cytb6*, which feeds them into PS I.

The water-splitting Complex

The step $H2O \rightarrow P680$ is performed by a poorly understood structure embedded within PS II called the *water-splitting complex* or the *oxygen-evolving complex*. It catalyzes a reaction that splits water into electrons, protons and oxygen:

$$2H_2O \rightarrow 4H^+ + 4e^- + O_2$$

The electrons are transferred to special chlorophyll molecules (embedded in PS II) that are promoted to a higher-energy state by the energy of photons.

The Reaction Center

The excitation $P680 \rightarrow P680^*$ of the reaction center pigment P680 occurs here. These special chlorophyll molecules embedded in PS II absorb the energy of photons, with maximal absorption at 680 nm. Electrons within these molecules are promoted to a higher-energy state. This is one of two core processes in photosynthesis, and it occurs with astonishing efficiency (greater than 90%) because, in addition to direct excitation by light at 680 nm, the energy of light first harvested by *antenna proteins* at other wavelengths in the light-harvesting system is also transferred to these special chlorophyll molecules.

This is followed by the step $P680^* \rightarrow$ pheophytin, and then on to plastoquinone, which occurs within the reaction center of PS II. High-energy electrons are transferred to plastoquinone before it subsequently picks up two protons to become plastoquinol. Plastoquinol is then released into the membrane as a mobile electron carrier.

This is the second core process in photosynthesis. The initial stages occur within *picosec-*

onds, with an efficiency of 100%. The seemingly impossible efficiency is due to the precise positioning of molecules within the reaction center. This is a solid-state process, not a chemical reaction. It occurs within an essentially crystalline environment created by the macromolecular structure of PS II. The usual rules of chemistry (which involve random collisions and random energy distributions) do not apply in solid-state environments.

Link of Water-splitting Complex and Chlorophyll Excitation

When the chlorophyll passes the electron to pheophytin, it obtains an electron from P_{680}^*. In turn, P_{680}^* can oxidize the Z (or Y_Z) molecule. Once oxidized, the Z molecule can derive electrons from the oxygen-evolving complex.

Summary

PS II is a transmembrane structure found in all chloroplasts. It splits water into electrons, protons and molecular oxygen. The electrons are transferred to plastoquinone, which carries them to a proton pump. Molecular oxygen is released into the atmosphere.

The emergence of such an incredibly complex structure, a macromolecule that converts the energy of sunlight into potentially useful work with efficiencies that are impossible in ordinary experience, seems almost magical at first glance. Thus, it is of considerable interest that, in essence, the same structure is found in purple bacteria.

Cytochrome b6

PS II and PS I are connected by a transmembrane proton pump, cytochrome *b* 6 complex (plastoquinol—plastocyanin reductase; EC 1.10.99.1). Electrons from PS II are carried by plastoquinol to cyt*b6*, where they are removed in a stepwise fashion (reforming plastoquinone) and transferred to a water-soluble electron carrier called *plastocyanin*. This redox process is coupled to the pumping of four protons across the membrane. The resulting proton gradient (together with the proton gradient produced by the water-splitting complex in PS II) is used to make ATP via ATP synthase.

The similarity in structure and function between cytochrome *b6* (in chloroplasts) and cytochrome*bc1* (*Complex III* in mitochondria) is striking. Both are transmembrane structures that remove electrons from a mobile, lipid-soluble electron carrier (plastoquinone in chloroplasts; ubiquinone in mitochondria) and transfer them to a mobile, water-soluble electron carrier (plastocyanin in chloroplasts; cytochrome *c* in mitochondria). Both are proton pumps that produce a transmembrane proton gradient.

Photosystem I

The cyclic light-dependent reactions occur when only the sole photosystem being used is photosystem 1. Photosystem 1 excites electrons which then cycle from the transport protein, ferredoxin (Fd), to the cytochrome complex, b6f, to another transport protein, plastocyanin (Pc), and back to photosystem I. A proton gradient is created across the thylakoid membrane (6) as protons (3) are transported from the chloroplast stroma (4) to the thylakoid lumen (5). Through chemiosmosis, ATP (9) is produced where ATP synthase (1) binds an inorganic phosphate group (8) to an ADP molecule (7).

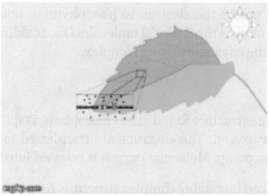

PS I accepts electrons from plastocyanin and transfers them either to NADPH (noncyclic electron transport) or back to cytochrome b6 (cyclic electron transport):

```
plastocyanin → P700 → P700* → FNR → NADPH

        ↑                        ↓

        b⁶          ←        plastoquinone
```

PS I, like PS II, is a complex, highly organized transmembrane structure that contains antenna chlorophylls, a reaction center (P700), phylloquinine, and a number of iron-sulfur proteins that serve as intermediate redox carriers.

The light-harvesting system of PS I uses multiple copies of the same transmembrane proteins used by PS II. The energy of absorbed light (in the form of delocalized, high-energy electrons) is funneled into the reaction center, where it excites special chlorophyll molecules (P700, maximum light absorption at 700 nm) to a higher energy level. The process occurs with astonishingly high efficiency.

Electrons are removed from excited chlorophyll molecules and transferred through a series of intermediate carriers to *ferredoxin*, a water-soluble electron carrier. As in PS II, this is a solid-state process that operates with 100% efficiency.

There are two different pathways of electron transport in PS I. In *noncyclic electron transport*, ferredoxin carries the electron to the enzyme ferredoxin NADP+

oxidoreductase (FNR) that reduces NADP+

to NADPH. In *cyclic electron transport*, electrons from ferredoxin are transferred (via plastoquinone) to a proton pump, cytochrome *b6*. They are then returned (via plasto-cyanin) to P700.

NADPH and ATP are used to synthesize organic molecules from CO_2. The ratio of NADPH to ATP production can be adjusted by adjusting the balance between cyclic and noncyclic electron transport.

It is noteworthy that PS I closely resembles photosynthetic structures found in green sulfur bacteria, just as PS II resembles structures found in purple bacteria.

Photosynthetic Electron Transport Chains in Bacteria

PS II, PS I, and cytochrome*b6* are found in chloroplasts. All plants and all photosynthetic algae contain chloroplasts, which produce NADPH and ATP by the mechanisms described above. In essence, the same transmembrane structures are also found in *cyanobacteria*.

Unlike plants and algae, cyanobacteria are prokaryotes. They do not contain chloroplasts. Rather, they bear a striking resemblance to chloroplasts themselves. This suggests that organisms resembling cyanobacteria were the evolutionary precursors of chloroplasts. One imagines primitive eukaryotic cells taking up cyanobacteria as intracellular symbionts.

Cyanobacteria

Cyanobacteria contain structures similar to PS II and PS I in chloroplasts. Their light-harvesting system is different from that found in plants (they use *phycobilins*, rather than chlorophylls, as antenna pigments), but their electron transport chain

$$H_2O \rightarrow PS\ II \rightarrow plastoquinone \rightarrow b_6 \rightarrow cytochrome\ c_6 \rightarrow PS\ I \rightarrow ferredoxin \rightarrow NADPH$$

$$\uparrow \qquad\qquad\qquad \downarrow$$

$$b_6 \qquad\qquad \leftarrow plastoquinone$$

is, in essence, the same as the electron transport chain in chloroplasts. The mobile water-soluble electron carrier is cytochrome *c6* in cyanobacteria, plastocyanin in plants.

Cyanobacteria can also synthesize ATP by oxidative phosphorylation, in the manner of other bacteria. The electron transport chain is

$$NADH\ dehydrogenase \rightarrow plastoquinone \rightarrow b_6 \rightarrow cytochrome\ c_6 \rightarrow cytochrome\ aa_3 \rightarrow O_2$$

where the mobile electron carriers are plastoquinone and cytochrome *c6*, while the proton pumps are NADH dehydrogenase, *b6* and cytochrome *aa3*.

Cyanobacteria are the only bacteria that produce oxygen during photosynthesis. Earth's primordial atmosphere was anoxic. Organisms like cyanobacteria produced our present-day oxygen-containing atmosphere.

The other two major groups of photosynthetic bacteria, purple bacteria and green sulfur bacteria, contain only a single photosystem and do not produce oxygen.

Purple Bacteria

Purple bacteria contain a single photosystem that is structurally related to PS II in cyanobacteria and chloroplasts:

$$P870 \rightarrow P870^* \rightarrow \text{ubiquinone} \rightarrow bc_1 \rightarrow \text{cytochrome } c_2 \rightarrow P870$$

This is a *cyclic* process in which electrons are removed from an excited chlorophyll molecule (*bacteriochlorophyll*; P870), passed through an electron transport chain to a proton pump (cytochrome *bc1* complex, similar but not identical to cytochrome *bc1* in chloroplasts), and then returned to the cholorophyll molecule. The result is a proton gradient, which is used to make ATP via ATP synthase. As in cyanobacteria and chloroplasts, this is a solid-state process that depends on the precise orientation of various functional groups within a complex transmembrane macromolecular structure.

To make NADPH, purple bacteria use an external electron donor (hydrogen, hydrogen sulfide, sulfur, sulfite, or organic molecules such as succinate and lactate) to feed electrons into a reverse electron transport chain.

Green Sulfur Bacteria

Green sulfur bacteria contain a photosystem that is analogous to PS I in chloroplasts:

$$P840 \rightarrow P840^* \rightarrow \text{ferredoxin} \rightarrow \text{NADH}$$

$$\uparrow \qquad\qquad\qquad\qquad\qquad \downarrow$$

$$\text{cyt } c_{553} \leftarrow bc_1 \leftarrow \text{menaquinone}$$

There are two pathways of electron transfer. In *cyclic electron transfer*, electrons are removed from an excited chlorophyll molecule, passed through an electron transport chain to a proton pump, and then returned to the chlorophyll. The mobile electron carriers are, as usual, a lipid-soluble quinone and a water-soluble cytochrome. The resulting proton gradient is used to make ATP.

In *noncyclic electron transfer*, electrons are removed from an excited chlorophyll molecule and used to reduce NAD^+ to NADH. The electrons removed from P840 must be replaced. This is accomplished by removing electrons from H2S, which is oxidized to sulfur (hence the name "green *sulfur* bacteria").

Purple bacteria and green sulfur bacteria occupy relatively minor ecological niches in the present day biosphere. They are of interest because of their importance in precambrian ecologies, and because their methods of photosynthesis were the likely evolutionary precursors of those in modern plants.

History

The first ideas about light being used in photosynthesis were proposed by Colin Flannery in 1779 who recognized it was sunlight falling on plants that was required, although Joseph Priestley had noted the production of oxygen without the association with light in 1772. Cornelius Van Niel proposed in 1931 that photosynthesis is a case of general mechanism where a photon of light is used to photo decompose a hydrogen donor and the hydrogen being used to reduce CO2. Then in 1939, Robin Hill showed that isolated chloroplasts would make oxygen, but not fix CO2 showing the light and dark reactions occurred in different places. Although they are referred to as light and dark reactions, both of them take place only in the presence of light. This led later to the discovery of photosystems 1 and 2.

Light-independent Reactions

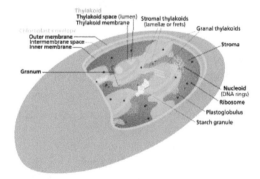

The internal structure of a chloroplast

The light-independent reactions of photosynthesis are chemical reactions that convert carbon dioxide and other compounds into glucose. These reactions occur in the stroma, the fluid-filled area of a chloroplast outside of the thylakoid membranes. These reactions take the products (ATP and NADPH) of light-dependent reactions and perform further chemical processes on them. There are three phases to the light-independent reactions, collectively called the Calvin cycle: carbon fixation, reduction reactions, and ribulose 1,5-bisphosphate (RuBP) regeneration.

Despite its name, this process occurs only when light is available. Plants do not carry out the Calvin cycle during nighttime. They instead release sucrose into the phloem from their starch reserves. This process happens when light is available independent of the kind of photosynthesis (C3 carbon fixation, C4 carbon fixation, and Crassulacean acid metabolism); CAM plants store malic acid in their vacuoles every night and release it by day in order to make this process work.

Coupling to Other Metabolic Pathways

These reactions are closely coupled to the thylakoid electron transport chain as reducing power provided by NADPH produced in the photosystem I is actively needed. The process of photorespiration, also known as C2 cycle, is also coupled to the dark reactions, as it results from an alternative reaction of the Rubisco enzyme, and its final byproduct is also another glyceraldehyde-3-P.

The Calvin cycle, Calvin–Benson–Bassham (CBB) cycle, reductive pentose phosphate cycle or C3 cycle is a series of biochemical redox reactions that take place in the stroma of chloroplast in photosynthetic organisms. It is also known as the light-independent reactions.

The cycle was discovered by Melvin Calvin, James Bassham, and Andrew Benson at the University of California, Berkeley by using the radioactive isotope carbon-14. It is one of the light-independent reactions used for carbon fixation.

Photosynthesis occurs in two stages in a cell. In the first stage, light-dependent reactions capture the energy of light and use it to make the energy-storage and transport molecules ATP and NADPH. The light-independent Calvin cycle uses the energy from short-lived electronically excited carriers to convert carbon dioxide and water into organic compounds that can be used by the organism (and by animals that feed on it). This set of reactions is also called *carbon fixation*. The key enzyme of the cycle is called RuBisCO. In the following biochemical equations, the chemical species (phosphates and carboxylic acids) exist in equilibria among their various ionized states as governed by the pH.

The enzymes in the Calvin cycle are functionally equivalent to most enzymes used in other metabolic pathways such as gluconeogenesis and the pentose phosphate pathway, but they are to be found in the chloroplast stroma instead of the cell cytosol, separating the reactions. They are activated in the light (which is why the name "dark reaction" is misleading), and also by products of the light-dependent reaction. These regulatory functions prevent the Calvin cycle from being respired to carbon dioxide. Energy (in the form of ATP) would be wasted in carrying out these reactions that have no net productivity.

The sum of reactions in the Calvin cycle is the following:

$$3 \text{ CO}_2 + 6 \text{ NADPH} + 5 \text{ H}_2\text{O} + 9 \text{ ATP} \rightarrow \text{glyceraldehyde-3-phosphate (G3P)} + 2 \text{ H}^+ + 6 \text{ NADP}^+ + 9 \text{ ADP} + 8 \text{ P}_i \quad (\text{P}_i = \text{inorganic phosphate})$$

Hexose (six-carbon) sugars are not a product of the Calvin cycle. Although many texts list a product of photosynthesis as C6H12O6, this is mainly a convenience to counter the equation of respiration, where six-carbon sugars are oxidized in mitochondria. The carbohydrate products of the Calvin cycle are three-carbon sugar phosphate molecules, or "triose phosphates," namely, glyceraldehyde-3-phosphate (G3P).

Steps

In the first stage of the Calvin cycle, a CO_2 molecule is incorporated into one of two three-carbon molecules (glyceraldehyde 3-phosphate or G3P), where it uses up two molecules of ATP and two molecules of NADPH, which had been produced in the light-dependent stage. The three steps involved are:

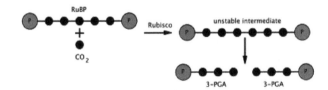

Calvin cycle step 1 (black circles represent carbon atoms)

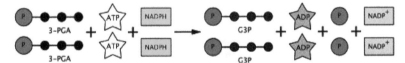

Calvin cycle steps 2 and 3 combined

1. The enzyme RuBisCO catalyses the carboxylation of ribulose-1,5-bisphosphate, RuBP, a 5-carbon compound, by carbon dioxide (a total of 6 carbons) in a two-step reaction. The product of the first step is enediol-enzyme complex that can capture CO2 or O2. Thus, enediol-enzyme complex is the real carboxylase/oxygenase. The CO2 that is captured by enediol in second step produces a six-carbon intermediate initially that immediately splits in half, forming two molecules of 3-phosphoglycerate, or 3-PGA, a 3-carbon compound (also: 3-phosphoglyceric acid, PGA, 3PGA).

2. The enzyme phosphoglycerate kinase catalyses the phosphorylation of 3-PGA by ATP (which was produced in the light-dependent stage). 1,3-Bisphosphoglycerate (1,3BPGA, glycerate-1,3-bisphosphate) and ADP are the products. (However, note that two 3-PGAs are produced for every CO2 that enters the cycle, so this step utilizes two ATP per CO2 fixed.)

3. The enzyme glyceraldehyde 3-phosphate dehydrogenase catalyses the reduction of 1,3BPGA by NADPH (which is another product of the light-dependent stage). Glyceraldehyde 3-phosphate (also called G3P, GP, TP, PGAL, GAP) is produced, and the NADPH itself is oxidized and becomes $NADP^+$. Again, two NADPH are utilized per CO2 fixed.

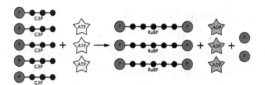

Regeneration stage of the Calvin cycle

The next stage in the Calvin cycle is to regenerate RuBP. Five G3P molecules produce three RuBP molecules, using up three molecules of ATP. Since each CO_2 molecule produces two G3P molecules, three CO_2 molecules produce six G3P molecules, of which five are used to regenerate RuBP, leaving a net gain of one G3P molecule per three CO_2 molecules (as would be expected from the number of carbon atoms involved).

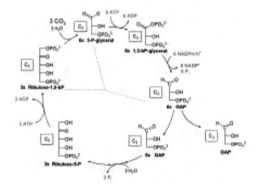

Simplified C3 cycle with structural formulas

The regeneration stage can be broken down into steps.

1. Triose phosphate isomerase converts all of the G3P reversibly into dihydroxyacetone phosphate (DHAP), also a 3-carbon molecule.

2. Aldolase and fructose-1,6-bisphosphatase convert a G3P and a DHAP into fructose 6-phosphate (6C). A phosphate ion is lost into solution.

3. Then fixation of another CO2 generates two more G3P.

4. F6P has two carbons removed by transketolase, giving erythrose-4-phosphate. The two carbons on transketolase are added to a G3P, giving the ketose xylulose-5-phosphate (Xu5P).

5. E4P and a DHAP (formed from one of the G3P from the second CO2 fixation) are converted into sedoheptulose-1,7-bisphosphate (7C) by aldolase enzyme.

6. Sedoheptulose-1,7-bisphosphatase (one of only three enzymes of the Calvin cycle that are unique to plants) cleaves sedoheptulose-1,7-bisphosphate into sedoheptulose-7-phosphate, releasing an inorganic phosphate ion into solution.

7. Fixation of a third CO2 generates two more G3P. The ketose S7P has two car-

bons removed by transketolase, giving ribose-5-phosphate (R5P), and the two carbons remaining on transketolase are transferred to one of the G3P, giving another Xu5P. This leaves one G3P as the product of fixation of 3 CO2, with generation of three pentoses that can be converted to Ru5P.

8. R5P is converted into ribulose-5-phosphate (Ru5P, RuP) by phosphopentose isomerase. Xu5P is converted into RuP by phosphopentose epimerase.

9. Finally, phosphoribulokinase (another plant-unique enzyme of the pathway) phosphorylates RuP into RuBP, ribulose-1,5-bisphosphate, completing the Calvin *cycle*. This requires the input of one ATP.

Thus, of six G3P produced, five are used to make three RuBP (5C) molecules (totaling 15 carbons), with only one G3P available for subsequent conversion to hexose. This requires nine ATP molecules and six NADPH molecules per three CO2 molecules. The equation of the overall Calvin cycle is shown diagrammatically below.

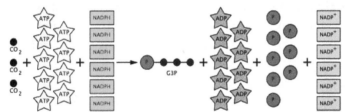

The overall equation of the Calvin cycle (black circles represent carbon atoms)

RuBisCO also reacts competitively with O2 instead of CO2 in photorespiration. The rate of photorespiration is higher at high temperatures. Photorespiration turns RuBP into 3-PGA and 2-phosphoglycolate, a 2-carbon molecule that can be converted via glycolate and glyoxalate to glycine. Via the glycine cleavage system and tetrahydrofolate, two glycines are converted into serine +CO2. Serine can be converted back to 3-phosphoglycerate. Thus, only 3 of 4 carbons from two phosphoglycolates can be converted back to 3-PGA. It can be seen that photorespiration has very negative consequences for the plant, because, rather than fixing CO2, this process leads to loss of CO2. C4 carbon fixation evolved to circumvent photorespiration, but can occur only in certain plants native to very warm or tropical climates—corn, for example.

Products

The immediate products of one turn of the Calvin cycle are 2 glyceraldehyde-3-phosphate (G3P) molecules, 3 ADP, and 2 NADP$^+$. (ADP and NADP$^+$ are not really "products." They are regenerated and later used again in the Light-dependent reactions). Each G3P molecule is composed of 3 carbons. In order for the Calvin cycle to continue, RuBP (ribulose 1,5-bisphosphate) must be regenerated. So, 5 out of 6 carbons from the 2 G3P molecules are used for this purpose. Therefore, there is only 1 net carbon produced to play with for each turn. To create 1 surplus G3P requires 3 carbons, and therefore 3 turns of the Calvin cycle. To make one glucose molecule (which can be created from 2 G3P molecules) would require 6 turns of the Calvin cycle. Surplus G3P can also be used to form other carbohydrates such as starch, sucrose, and cellulose, depending on what the plant needs.

Light-dependent Regulation

Despite its widespread names (both light-independent and dark reactions), these reactions do not occur in the dark or at night. There is a light-dependent regulation of the cycle enzymes, as the third step requires reduced NADP; and this process would be a waste of energy, as there is no electron flow in the dark.

There are 2 regulation systems at work when the cycle needs to be turned on or off: thioredoxin/ferredoxin activation system, which activates some of the cycle enzymes, and the RuBisCo enzyme activation, active in the Calvin cycle, which involves its own activase.

The thioredoxin/ferredoxin system activates the enzymes glyceraldehyde-3-P dehydrogenase, glyceraldehyde-3-P phosphatase, fructose-1,6-bisphosphatase, sedoheptulose-1,7-bisphosphatase, and ribulose-5-phosphatase kinase, which are key points of the process. This happens when light is available, as the ferredoxin protein is reduced in the photosystem I complex of the thylakoid electron chain when electrons are circulating through it. Ferredoxin then binds to and reduces the thioredoxin protein, which activates the cycle enzymes by severing a cystine bond found in all these enzymes. This is a dynamic process as the same bond is formed again by other proteins that deactivate the enzymes. The implications of this process are that the enzymes remain mostly activated by day and are deactivated in the dark when there is no more reduced ferredoxin available.

The enzyme RuBisCo has its own activation process, which involves a more complex process. It is necessary that a specific lysine amino acid be carbamylated in order to activate the enzyme. This lysine binds to RuBP and leads to a non-functional state if left uncarbamylated. A specific activase enzyme, called RuBisCo activase, helps this carbamylation process by removing one proton from the lysine and making the binding of the carbon dioxide molecule possible. Even then the RuBisCo enzyme is not yet functional, as it needs a magnesium ion to be bound to the lysine in order to function. This magnesium ion is released from the thylakoid lumen when the inner pH drops due to the active pumping of protons from the electron flow. RuBisCo activase itself is activated by increased concentrations of ATP in the stroma caused by its phosphorylation.

C_4 Carbon Fixation

C_4 carbon fixation is one of three biochemical processes, along with C_3 and CAM photosynthesis, that fixes carbon. It is named for the 4-carbon molecule of the first product of carbon fixation found in the small subset of plants that use the C_4 process. This process is in contrast to the 3-carbon molecule products of C_3 plants.

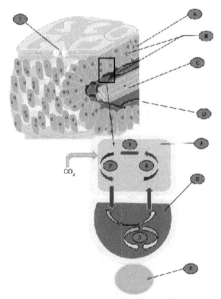

Kranz anatomy allows C4 carbon fixation.
A: Mesophyll Cell
B: Chloroplast
C: Vascular Tissue
D: Bundle Sheath Cell
E: Stroma

F: Vascular Tissue: provides continuous source of water
1) Carbon is fixed to produce oxaloacetate by PEP carboxylase.
2) The four carbon molecule then exits the cell and enters the chloroplasts of bundle sheath cells.
3) It is then broken down releasing carbon dioxide and producing pyruvate. Carbon dioxide combines with ribulose bisphosphate and proceeds to the Calvin Cycle.
4) Pyruvate re-enters the mesophyll cell. It then reacts with ATP to produce the beginning compound of the C4 cycle.

C_4 fixation is an elaboration of the more common C_3 carbon fixation and is believed to have evolved more recently. C_4 and CAM overcome the tendency of the enzyme RuBisCO to wastefully fix oxygen rather than carbon dioxide in the process of photorespiration. This is achieved in a more efficient environment for RubisCo by shuttling CO_2 via malate or aspartate from mesophyll cells to bundle-sheath cells. In these bundle-sheath cells, RuBisCO is isolated from atmospheric oxygen and saturated with the CO_2 released by decarboxylation of the malate. C4 plants use PEP carboxylase to capture more CO_2 in the mesophyll cells. PEP Carboxylase (3 carbons) binds to CO_2 to make oxaloacetic acid (OAA). The OAA then makes malate (4 carbons). Malate enters bundle sheath cells and releases the CO_2 where RuBisCO works more efficiently. These additional steps, however, require more energy in the form of ATP. Because of this extra energy requirement, C_4 plants are able to more efficiently fix carbon in drought, high temperatures, and limitations of nitrogen or CO_2, while the more common C_3 pathway is more efficient in the other conditions.

C_4 Pathway

The first experiments indicating that some plants do not use C3 carbon fixation but instead produce malate and aspartate in the first step of carbon fixation were done in the 1950s and early 1960s by Hugo P. Kortschak and Yuri Karpilov. The C_4 pathway was elucidated by Marshall Davidson Hatch and C. R. Slack, in Australia, in 1966; it is sometimes called the Hatch-Slack pathway.

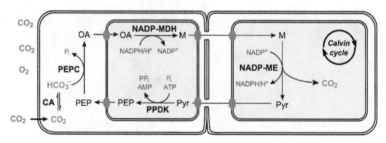

NADP-ME type C_4 pathway

In C_3 plants, the first step in the light-independent reactions of photosynthesis involves the fixation of CO_2 by the enzyme RuBisCO into 3-phosphoglycerate. However, due to the dual carboxylase and oxygenase activity of RuBisCo, some part of the substrate is oxidized rather than carboxylated, resulting in loss of substrate and consumption of energy, in what is known as photorespiration. In order to bypass the photorespiration pathway, C_4 plants have developed a mechanism to efficiently deliver CO_2 to the RuBis-CO enzyme. They utilize their specific leaf anatomy where chloroplasts exist not only in the mesophyll cells in the outer part of their leaves but in the bundle sheath cells as well. Instead of direct fixation to RuBisCO in the Calvin cycle, CO_2 is incorporated into a 4-carbon organic acid, which has the ability to regenerate CO_2 in the chloroplasts of the bundle sheath cells. Bundle sheath cells can then utilize this CO_2 to generate carbohydrates by the conventional C_3 pathway.

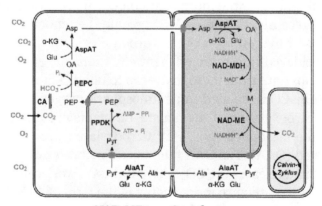

NAD-ME type C_4 pathway

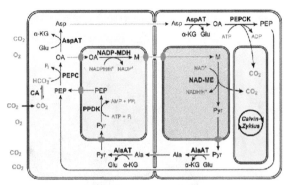

PEPCK type C_4 pathway

The first step in the pathway is the conversion of pyruvate to phosphoenolpyruvate (PEP), by the enzyme pyruvate orthophosphate dikinase. This reaction requires inorganic phosphate and ATP plus pyruvate, producing phosphoenolpyruvate, AMP, and inorganic pyrophosphate (PPi). The next step is the fixation of CO_2 into oxaloacetate by the enzyme PEP carboxylase. Both of these steps occur in the mesophyll cells:

$$pyruvate + P_i + ATP \rightarrow PEP + AMP + PP_i \quad PEP + CO_2 \rightarrow oxaloacetate$$

PEP carboxylase has a lower Km for HCO_3^- — and, hence, higher affinity — than RuBisCO. Furthermore, O_2 is a very poor substrate for this enzyme. Thus, at relatively low concentrations of CO_2, most CO_2 will be fixed by this pathway.

The product is usually converted to malate, a simple organic compound, which is transported to the bundle-sheath cells surrounding a nearby vein. Here, it is decarboxylated to produce CO_2 and pyruvate. The CO_2 now enters the Calvin cycle and the pyruvate is transported back to the mesophyll cell.

Since every CO_2 molecule has to be fixed twice, first by 4-carbon organic acid and second by RuBisCO, the C_4 pathway uses more energy than the C_3 pathway. The C_3 pathway requires 18 molecules of ATP for the synthesis of one molecule of glucose, whereas the C_4 pathway requires 30 molecules of ATP. This energy debt is more than paid for by avoiding losing more than half of photosynthetic carbon in photorespiration as occurs in some tropical plants, making it an adaptive mechanism for minimizing the loss.

There are several variants of this pathway:

1. The 4-carbon acid transported from mesophyll cells may be malate, as above, or aspartate

2. The 3-carbon acid transported back from bundle-sheath cells may be pyruvate, as above, or alanine

3. The enzyme that catalyses decarboxylation in bundle-sheath cells differs. In maize and sugarcane, the enzyme is NADP-malic enzyme; in millet, it is NAD-malic enzyme; and, in *Panicum maximum*, it is PEP carboxykinase.

C_4 Leaf Anatomy

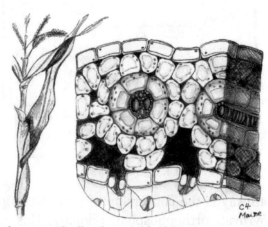

Cross section of a C4 plant, specifically of a maize leaf. Kranz anatomy shown. Drawing based on microscopic images courtesy of Cambridge University Plant Sciences Department.

The C_4 plants often possess a characteristic leaf anatomy called *kranz anatomy*, from the German word for wreath. Their vascular bundles are surrounded by two rings of cells; the inner ring, called bundle sheath cells, contains starch-rich chloroplasts lacking grana, which differ from those in mesophyll cells present as the outer ring. Hence, the chloroplasts are called dimorphic. The primary function of kranz anatomy is to provide a site in which CO_2 can be concentrated around RuBisCO, thereby avoiding photorespiration. In order to maintain a significantly higher CO_2 concentration in the bundle sheath compared to the mesophyll, the boundary layer of the kranz has a low conductance to CO_2, a property that may be enhanced by the presence of suberin.

Although most C_4 plants exhibit kranz anatomy, there are, however, a few species that operate a limited C_4 cycle without any distinct bundle sheath tissue. *Suaeda aralocaspica, Bienertia cycloptera, Bienertia sinuspersici* and *Bienertia kavirense* (all chenopods) are terrestrial plants that inhabit dry, salty depressions in the deserts of the Middle East. These plants have been shown to operate single-cell C_4 CO_2-concentrating mechanisms, which are unique among the known C_4 mechanisms. Although the cytology of both genera differs slightly, the basic principle is that fluid-filled vacuoles are employed to divide the cell into two separate areas. Carboxylation enzymes in the cytosol can, therefore, be kept separate from decarboxylase enzymes and RuBisCO in the chloroplasts, and a diffusive barrier can be established between the chloroplasts (which contain RuBisCO) and the cytosol. This enables a bundle-sheath-type area and a mesophyll-type area to be established within a single cell. Although this does allow a limited C_3 cycle to operate, it is relatively inefficient, with the occurrence of much leakage of CO_2 from around RuBisCO. There is also evidence for the exhibiting of inducible C_4 photosynthesis by non-kranz aquatic macrophyte *Hydrilla verticillata* under warm conditions, although the mechanism by which CO_2 leakage from around RuBisCO is minimised is currently uncertain.

The Evolution and Advantages of the C_4 Pathway

C_4 plants have a competitive advantage over plants possessing the more common C_3 carbon fixation pathway under conditions of drought, high temperatures, and nitrogen or CO_2 limitation. When grown in the same environment, at 30 °C, C_3 grasses lose approximately 833 molecules of water per CO_2 molecule that is fixed, whereas C_4 grasses lose only 277. This increased water use efficiency of C_4 grasses means that soil moisture is conserved, allowing them to grow for longer in arid environments.

C_4 carbon fixation has evolved on up to 40 independent occasions in different families of plants, making it a prime example of convergent evolution. This convergence may have been facilitated by the fact that many potential evolutionary pathways to a C_4 phenotype exist, many of which involve initial evolutionary steps not directly related to photosynthesis. C_4 plants arose around 25 to 32 million years ago during the Oligocene (precisely when is difficult to determine) and did not become ecologically significant until around 6 to 7 million years ago, in the Miocene Period. C_4 metabolism originated when grasses migrated from the shady forest undercanopy to more open environments, where the high sunlight gave it an advantage over the C_3 pathway. Drought was not necessary for its innovation; rather, the increased resistance to water stress was a by-product of the pathway and allowed C_4 plants to more readily colonise arid environments.

Today, C_4 plants represent about 5% of Earth's plant biomass and 3% of its known plant species. Despite this scarcity, they account for about 30% of terrestrial carbon fixation. Increasing the proportion of C_4 plants on earth could assist biosequestration of CO_2 and represent an important climate change avoidance strategy. Present-day C_4 plants are concentrated in the tropics and subtropics (below latitudes of 45°) where the high air temperature contributes to higher possible levels of oxygenase activity by RuBisCO, which increases rates of photorespiration in C_3 plants.

Plants that Use C_4 Carbon Fixation

About 7,600 plant species use C_4 carbon fixation, which represents about 3% of all terrestrial species of plants. All these 7,600 species are angiosperms. C_4 carbon fixation is less common in dicots than in monocots, with only 4.5% of dicots using the C_4 pathway, compared to 40% of monocots. Despite this, only three families of monocots utilise C_4 carbon fixation compared to 15 dicot families. Of the monocot clades containing C_4 plants, the grass (Poaceae) species use the C_4 photosynthetic pathway most. Forty-six percent of grasses are C_4 and together account for 61% of C_4 species. These include the food crops maize, sugar cane, millet, and sorghum. Of the dicot clades containing C_4 species, the order Caryophyllales contains the most species. Of the families in the Caryophyllales, the Chenopodiaceae use C_4 carbon fixation the most, with 550 out of 1,400 species using it. About 250 of the 1000 species of the related Amaranthaceae also use C_4.

Members of the sedge family Cyperaceae, and numerous families of Eudicots, including the daisies Asteraceae, cabbages Brassicaceae, and spurges Euphorbiaceae also use C_4.

Converting C_3 Plants to C_4

Given the advantages of C_4, a group of scientists from institutions around the world are working on the C_4 Rice Project to turn rice, a C_3 plant, into a C_4 plant by studying the C4 plants corn and *Brachypodium*. As rice is the world's most important human food—it is the staple food for more than half the planet—having rice that is more efficient at converting sunlight into grain could have significant global benefits towards improving food security. The team claim C_4 rice could produce up to 50% more grain—and be able to do it with less water and nutrients.

The researchers have already identified genes needed for C_4 photosynthesis in rice and are now looking towards developing a prototype C_4 rice plant. In 2012, the Government of the United Kingdom along with the Bill & Melinda Gates Foundation provided $14 million over 3 years towards the C_4 Rice Project at the International Rice Research Institute.

Photodissociation

Photodissociation, photolysis, or photodecomposition is a chemical reaction in which a chemical compound is broken down by photons. It is defined as the interaction of one or more photons with one target molecule. Photodissociation is not limited to visible light. Any photon with sufficient energy can affect the chemical bonds of a chemical compound. Since a photon's energy is inversely proportional to its wavelength, electromagnetic waves with the energy of visible light or higher, such as ultraviolet light, x-rays and gamma rays are usually involved in such reactions.

Photolysis in Photosynthesis

Photolysis is part of the light-dependent reactions of photosynthesis. The general reaction of photosynthetic photolysis can be given as

$$H_2A + 2 \text{ photons (light)} \rightarrow 2 \text{ e}^- + 2 \text{ H}^+ + A$$

The chemical nature of "A" depends on the type of organism. In purple sulfur bacteria, hydrogen sulfide (H_2S) is oxidized to sulfur (S). In oxygenic photosynthesis, water (H_2O) serves as a substrate for photolysis resulting in the generation of diatomic oxygen (O_2). This is the process which returns oxygen to Earth's atmosphere. Photolysis of water occurs in the thylakoids of cyanobacteria and the chloroplasts of green algae and plants.

Energy Transfer Models

The conventional, semi-classical, model describes the photosynthetic energy transfer process as one in which excitation energy hops from light-capturing pigment molecules to reaction center molecules step-by-step down the molecular energy ladder.

The effectiveness of photons of different wavelengths depends on the absorption spectra of the photosynthetic pigments in the organism. Chlorophylls absorb light in the violet-blue and red parts of the spectrum, while accessory pigments capture other wavelengths as well. The phycobilins of red algae absorb blue-green light which penetrates deeper into water than red light, enabling them to photosynthesize in deep waters. Each absorbed photon causes the formation of an exciton (an electron excited to a higher energy state) in the pigment molecule. The energy of the exciton is transferred to a chlorophyll molecule (P680, where P stands for pigment and 680 for its absorption maximum at 680 nm) in the reaction center of photosystem II via resonance energy transfer. P680 can also directly absorb a photon at a suitable wavelength.

Photolysis during photosynthesis occurs in a series of light-driven oxidation events. The energized electron (exciton) of P680 is captured by a primary electron acceptor of the photosynthetic electron transfer chain and thus exits photosystem II. In order to repeat the reaction, the electron in the reaction center needs to be replenished. This occurs by oxidation of water in the case of oxygenic photosynthesis. The electron-deficient reaction center of photosystem II (P680*) is the strongest biological oxidizing agent yet discovered, which allows it to break apart molecules as stable as water.

The water-splitting reaction is catalyzed by the oxygen evolving complex of photosystem II. This protein-bound inorganic complex contains four manganese ions, plus calcium and chloride ions as cofactors. Two water molecules are complexed by the manganese cluster, which then undergoes a series of four electron removals (oxidations) to replenish the reaction center of photosystem II. At the end of this cycle, free oxygen (O_2) is generated and the hydrogen of the water molecules has been converted to four protons released into the thylakoid lumen.

These protons, as well as additional protons pumped across the thylakoid membrane coupled with the electron transfer chain, form a proton gradient across the membrane that drives photophosphorylation and thus the generation of chemical energy in the form of adenosine triphosphate (ATP). The electrons reach the P700 reaction center of photosystem I where they are energized again by light. They are passed down another electron transfer chain and finally combine with the coenzyme $NADP^+$ and protons outside the thylakoids to NADPH. Thus, the net oxidation reaction of water photolysis can be written as:

$$2 \text{ H}_2\text{O} + 2 \text{ NADP}^+ + 8 \text{ photons (light)} \rightarrow 2 \text{ NADPH} + 2 \text{ H}^+ + \text{O}_2$$

The free energy change (ΔG) for this reaction is 102 kilocalories per mole. Since the energy of light at 700 nm is about 40 kilocalories per mole of photons, approximately 320 kilocalories of light energy are available for the reaction. Therefore, approximately one-third of the available light energy is captured as NADPH during photolysis and electron transfer. An equal amount of ATP is generated by the resulting proton gradi ent. Oxygen as a byproduct is of no further use to the reaction and thus released into the atmosphere.

Quantum Models

In 2007 a quantum model was proposed by Graham Fleming and his co-workers which includes the possibility that photosynthetic energy transfer might involve quantum oscillations, explaining its unusually high efficiency.

According to Fleming there is direct evidence that remarkably long-lived wavelike electronic quantum coherence plays an important part in energy transfer processes during photosynthesis, which can explain the extreme efficiency of the energy transfer because it enables the system to sample all the potential energy pathways, with low loss, and choose the most efficient one. This claim has, however, since been proven wrong in several publications .

This approach has been further investigated by Gregory Scholes and his team at the University of Toronto, which in early 2010 published research results that indicate that some marine algae make use of quantum-coherent electronic energy transfer (EET) to enhance the efficiency of their energy harnessing.

Photolysis in the Atmosphere

Photolysis occurs in the atmosphere as part of a series of reactions by which primary pollutants such as hydrocarbons and nitrogen oxides react to form secondary pollutants such as peroxyacyl nitrates.

The two most important photodissociaton reactions in the troposphere are firstly:

$$O_3 + h\nu \rightarrow O_2 + O(^1D) \quad \lambda < 320 \text{ nm}$$

which generates an excited oxygen atom which can react with water to give the hydroxyl radical:

$$O(^1D) + H_2O \rightarrow 2\ ^{\bullet}OH$$

The hydroxyl radical is central to atmospheric chemistry as it initiates the oxidation of hydrocarbons in the atmosphere and so acts as a detergent.

Secondly the reaction:

$$NO_2 + h\nu \rightarrow NO + O$$

is a key reaction in the formation of tropospheric ozone.

The formation of the ozone layer is also caused by photodissociation. Ozone in the Earth's stratosphere is created by ultraviolet light striking oxygen molecules containing two oxygen atoms (O_2), splitting them into individual oxygen atoms (atomic oxygen).

The atomic oxygen then combines with unbroken O_2 to create ozone, O_3. In addition, photolysis is the process by which CFCs are broken down in the upper atmosphere to form ozone-destroying chlorine free radicals.

Astrophysics

In astrophysics, photodissociation is one of the major processes through which molecules are broken down (but new molecules are being formed). Because of the vacuum of the interstellar medium, molecules and free radicals can exist for a long time. Photodissociation is the main path by which molecules are broken down. Photodissociation rates are important in the study of the composition of interstellar clouds in which stars are formed.

Examples of photodissociation in the interstellar medium are hv is the energy of a single photon of frequency v):

$$H_2O + hv \rightarrow H + OH$$

$$CH_4 + hv \rightarrow CH_3 + H$$

Atmospheric Gamma-ray Bursts

Currently orbiting satellites detect an average of about one gamma-ray burst per day. Because gamma-ray bursts are visible to distances encompassing most of the observable universe, a volume encompassing many billions of galaxies, this suggests that gamma-ray bursts must be exceedingly rare events per galaxy.

Measuring the exact rate of gamma-ray bursts is difficult, but for a galaxy of approximately the same size as the Milky Way, the expected rate (for long GRBs) is about one burst every 100,000 to 1,000,000 years. Only a few percent of these would be beamed towards Earth. Estimates of rates of short GRBs are even more uncertain because of the unknown beaming fraction, but are probably comparable.

A gamma-ray burst in the Milky Way, if close enough to Earth and beamed towards it, could have significant effects on the biosphere. The absorption of radiation in the atmosphere would cause photodissociation of nitrogen, generating nitric oxide that would act as a catalyst to destroy ozone.

The atmospheric photodissociation

- $N_2 \rightarrow 2N$

- $O_2 \rightarrow 2O$

- $CO_2 \rightarrow C + 2O$

- $H_2O \rightarrow 2H + O$

- $2NH_3 \rightarrow 3H_2 + N_2$

would yield

- NO_2 (consumes up to 400 ozone molecules)
- CH_2 (nominal)
- CH_4 (nominal)
- CO_2

According to a 2004 study, a GRB at a distance of about a kiloparsec could destroy up to half of Earth's ozone layer; the direct UV irradiation from the burst combined with additional solar UV radiation passing through the diminished ozone layer could then have potentially significant impacts on the food chain and potentially trigger a mass extinction. The authors estimate that one such burst is expected per billion years, and hypothesize that the Ordovician-Silurian extinction event could have been the result of such a burst.

There are strong indications that long gamma-ray bursts preferentially or exclusively occur in regions of low metallicity. Because the Milky Way has been metal-rich since before the Earth formed, this effect may diminish or even eliminate the possibility that a long gamma-ray burst has occurred within the Milky Way within the past billion years. No such metallicity biases are known for short gamma-ray bursts. Thus, depending on their local rate and beaming properties, the possibility for a nearby event to have had a large impact on Earth at some point in geological time may still be significant.

Multiple Photon Dissociation

Single photons in the infrared spectral range usually are not energetic enough for direct photodissociation of molecules. However, after absorption of multiple infrared photons a molecule may gain internal energy to overcome its barrier for dissociation. Multiple photon dissociation (MPD, IRMPD with infrared radiation) can be achieved by applying high power lasers, e.g. a carbon dioxide laser, or a free electron laser, or by long interaction times of the molecule with the radiation field without the possibility for rapid cooling, e.g. by collisions. The latter method allows even for MPD induced by black body radiation, a technique called blackbody infrared radiative dissociation (BIRD).

Oxygen Evolution

Oxygen evolution is the process of generating molecular oxygen through chemical reaction. Mechanisms of oxygen evolution include the oxidation of water during oxygenic photosynthesis, electrolysis of water into oxygen and hydrogen, and electrocatalytic oxygen evolution from oxides and oxoacids.

Photosynthetic oxygen evolution is the fundamental process by which breathable oxygen is generated in earth's biosphere. The reaction is part of the light-dependent reactions of photosynthesis in cyanobacteria and the chloroplasts of green algae and plants. It utilizes the energy of light to split a water molecule into its protons and electrons for photosynthesis. Free oxygen is generated as a by-product of this reaction, and is released into the atmosphere.

Biochemical Reaction

Photosynthetic oxygen evolution occurs via the light-dependent oxidation of water to molecular oxygen and can be written as the following simplified chemical reaction:

$$2H_2O \rightarrow 4e^- + 4H^+ + O_2$$

The reaction requires the energy of four photons. The electrons from the oxidized water molecules replace electrons in the P_{680} component of photosystem II that have been removed into an electron transport chain via light-dependent excitation and resonance energy transfer onto plastoquinone. Photosytem II, therefore, has also been referred to as water-plastoquinone oxido-reductase. The protons are released into the thylakoid lumen, thus contributing to the generation of a proton gradient across the thylakoid membrane. This proton gradient is the driving force for ATP synthesis via photophosphorylation and coupling the absorption of light energy and oxidation of water to the creation of chemical energy during photosynthesis.

Oxygen-evolving Complex

Water oxidation is catalyzed by a manganese-containing cofactor contained in photosystem II known as the oxygen-evolving complex (OEC) or water-splitting complex. Manganese is an important cofactor, and calcium and chloride are also required for the reaction to occur.

X-ray crystallographic data have been used to propose a structure and mechanism of action for the oxygen-evolving complex and its manganese cluster. Based on structural and spectroscopic experiments, oxygen evolution involves a core three-plus-one cluster of three manganese ions and one calcium ion, with one additional manganese, which are oxidized via intermediate states called *S-states*. The O-O bond of molecular oxygen is formed between manganese-ligated oxygen atoms at the most oxidized, or S4, state.

History of Discovery

It was not until the end of the 18th century that Joseph Priestley discovered by accident the ability of plants to "restore" air that had been "injured" by the burning of a candle. He followed up on the experiment by showing that air "restored" by vegetation was *"not at all inconvenient to a mouse."* He was later awarded a medal for his discoveries that: *"...no vegetable grows in vain... but cleanses and purifies our atmosphere."* Priestley's experiments were followed up by Jan Ingenhousz, a Dutch physician, who showed that "restoration" of air only worked in the presence of light and green plant parts.

Ingenhousz suggested in 1796 that CO_2 (carbon dioxide) is split during photosynthesis to release oxygen, while the carbon combined with water to form carbohydrates. While this hypothesis was attractive and reasonable and thus widely accepted for a long time, it was later proven incorrect. Graduate student C.B. Van Niel at Stanford University found that purple sulfur bacteria reduce carbon to carbohydrates, but accumulate sulfur instead of releasing oxygen. He boldly proposed that, in analogy to the sulfur bacteria's forming elemental sulfur from H_2S (hydrogen sulfide), plants would form oxygen from H_2O (water). In 1937, this hypothesis was corroborated by the discovery that plants are capable of producing oxygen in the absence of CO_2. This discovery was made by Robin Hill, and subsequently the light-driven release of oxygen in the absence of CO_2 was called the *Hill reaction*. Our current knowledge of the mechanism of oxygen evolution during photosynthesis was further established in experiments tracing isotopes of oxygen from water to oxygen gas.

Technological Oxygen Evolution

Oxygen evolution occurs as a byproduct of hydrogen production via electrolysis of water. While oxygen production is not the main focus of industrial applications of water electrolysis, it becomes essential for life support systems in situations that require the generation of oxygen for air revitalization. Human exploration of regions that lack breathable oxygen, such as the deep sea or outer space, requires means of reliably generating oxygen apart from earth's atmosphere. Submarines and spacecraft utilize either an electrolytic mechanism (water or solid oxide electrolysis) or chemical oxygen generators as part of their life support equipment.

References

- Campbell, Neil A.; Reece, Jane B. (2005). Biology (7th ed.). San Francisco: Pearson - Benjamin Cummings. pp. 186–191. ISBN 0-8053-7171-0.

- Raven, Peter H.; Ray F. Evert; Susan E. Eichhorn (2005). Biology of Plants (7th ed.). New York: W.H. Freeman and Company Publishers. pp. 115–127. ISBN 0-7167-1007-2.

- Campbell, Neil A.; Brad Williamson; Robin J. Heyden (2006). Biology: Exploring Life. Boston, Massachusetts: Pearson Prentice Hall. ISBN 0-13-250882-6.

- David Oakley Hall; K. K. Rao; Institute of Biology (1999). Photosynthesis. Cambridge University Press. ISBN 978-0-521-64497-6. Retrieved 3 November 2011.

- Gilles van Kote (2012-01-24). "Researchers aim to flick the high-carbon switch on rice". The Guardian. Retrieved 2012-11-10.

- N. Christenson; H. F. Kauffmann; T. Pullerits; T. Mancal (2012). "Origin of Long-Lived Coherences in Light-Harvesting Complexes". J. Phys. Chem. B. 116: 7449–7454.

Various Aspects of Photosynthesis

This chapter studies the various aspects of photosynthesis like photophosphorylation, photorespiration, photosynthate partitioning and photosynthetically active radiation. To understand these processes, the chapter introduces readers to concepts like chloroplast, thylakoid and PI curve. There is also a section dedicated to anoxygenic photosynthesis carried out by anaerobes like green sulphur bacteria, purple bacteria, acidobacteria and heliobacteria.

Anoxygenic Photosynthesis

Bacterial anoxygenic photosynthesis is distinguished from the more familiar terrestrial plant oxygenic photosynthesis by the nature of the terminal reductant (e.g. hydrogen sulfide rather than water) and in the byproduct generated (e.g. elemental sulfur instead of molecular oxygen). As its name implies, anoxygenic photosynthesis does not produce oxygen as a byproduct of the reaction. Additionally, all known organisms that carry out anoxygenic photosythesis are obligate anaerobes. Several groups of bacteria can conduct anoxygenic photosynthesis: green sulfur bacteria (GSB), red and green filamentous phototrophs (FAPs e.g. Chloroflexi), purple bacteria, Acidobacteria, and heliobacteria.

The pigments used to carry out anaerobic photosynthesis are similar to chlorophyll but differ in molecular detail and peak wavelength of light absorbed. Bacteriochlorophylls *a* through *g* absorb electromagnetic photons maximally in the near-infrared within their natural membrane milieu. This differs from chlorophyll a, the predominant plant and cyanobacteria pigment, which has peak absorption wavelength approximately 100 nanometers shorter (in the red portion of the visible spectrum).

Some archaea (e.g. *Halobacterium*) capture light energy for metabolic function and are thus phototrophic but none are known to "fix" carbon (i.e. be photosynthetic). Instead of a chlorophyll-type receptor and electron transport chain, proteins such as halorhodopsin capture light energy with the aid of diterpenes to move ions against the gradient and produce ATP via chemiosmosis in the manner of mitochondria.

There are two main types of anaerobic photosynthetic electron transport chains in bacteria. The type I reaction centers found in GSB, Chloracidobacterium, and Heliobacteria and the type II reaction centers found in FAPs and Purple Bacteria

Type I Reaction Centers

The electron transport chain of green sulfur bacteria — such as is present in model organism *Chlorobaculum tepidum* — uses the reaction centre bacteriochlorophyll pair, P840. When light is absorbed by the reaction center, P840 enters an excited state with a large negative reduction potential, and so readily donates the electron to bacteriochlorophyll 663 which passes it on down the electron chain. The electron is transferred through a series of electron carriers and complexes until it is used to reduce NAD$^+$. P840 regeneration is accomplished with the oxidation of sulfide ion from hydrogen sulfide (or hydrogen or ferrous iron) by cytochrome.

Type II Reaction Centers

Although the type II reaction centers are structurally and sequentially analogous to Photosystem II (PSII) in plant chloroplasts and cyanobacteria, known organisms that exhibit anoygenic photosynthesis do not have a region analogous to the oxygen-evolving complex of PSII.

The electron transport chain of purple non-sulfur bacteria begins when the reaction centre bacteriochlorophyll pair, P870, becomes excited from the absorption of light. Excited P870 will then donate an electron to bacteriopheophytin, which then passes it on to a series of electron carriers down the electron chain. In the process, it will generate an electro-chemical gradient which can then be used to synthesize ATP by chemiosmosis. P870 has to be regenerated (reduced) to be available again for a photon reaching the reaction-center to start the process anew. Molecular hydrogen in the bacterial environment is the usual electron donor.

Photophosphorylation

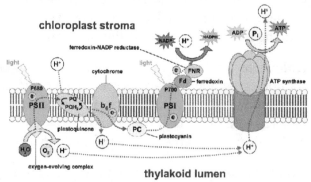

Noncyclic photophosphorylation through light-dependent reactions of photosynthesis at the thylakoid membrane

In the process of photosynthesis, the phosphorylation of ADP to form ATP using the energy of sunlight is called photophosphorylation. Only two sources of energy are available to living organisms: sunlight and reduction-oxidation (redox) reactions. All organisms produce ATP, which is the universal energy currency of life. This involves photolysis of water and a continuous unidirectional flow of electron from water to PS||.

In photophosphorylation, light energy is used to create a high-energy electron donor and a lower-energy electron acceptor. Electrons then move spontaneously from donor to acceptor through an electron transport chain.

ATP and Reactions

ATP is made by an enzyme called ATP synthase. Both the structure of this enzyme and its underlying gene are remarkably similar in all known forms of life.

ATP synthase is powered by a transmembrane electrochemical potential gradient, usually in the form of a proton gradient. The function of the electron transport chain is to produce this gradient. In all living organisms, a series of redox reactions is used to produce a trans-membrane electrochemical potential gradient, or a so-called proton motive force (pmf).

Redox reactions are chemical reactions in which electrons are transferred from a donor molecule to an acceptor molecule. The underlying force driving these reactions is the Gibbs free energy of the reactants and products. The Gibbs free energy is the energy available ("free") to do work. Any reaction that decreases the overall Gibbs free energy of a system will proceed spontaneously (given that the system is isobaric and also adiabatic), although the reaction may proceed slowly if it is kinetically inhibited.

The transfer of electrons from a high-energy molecule (the donor) to a lower-energy molecule (the acceptor) can be *spatially* separated into a series of intermediate redox reactions. This is an electron transport chain.

The fact that a reaction is thermodynamically possible does not mean that it will actually occur. A mixture of hydrogen gas and oxygen gas does not spontaneously ignite. It is necessary either to supply an activation energy or to lower the intrinsic activation energy of the system, in order to make most biochemical reactions proceed at a useful rate. Living systems use complex macromolecular structures to lower the activation energies of biochemical reactions.

It is possible to couple a thermodynamically favorable reaction (a transition from a high-energy state to a lower-energy state) to a thermodynamically unfavorable reaction (such as a separation of charges, or the creation of an osmotic gradient), in such a way that the overall free energy of the system decreases (making it thermodynamically possible), while useful work is done at the same time. The principle that biological macromolecules catalyze a thermodynamically unfavorable reaction *if and only if* a thermodynamically favorable reaction occurs simultaneously, underlies all known forms of life.

Electron transport chains (most known as ETC) produce energy in the form of a trans-membrane electrochemical potential gradient. This energy is used to do useful work. The gradient can be used to transport molecules across membranes. It can be used to do mechanical work, such as rotating bacterial flagella. It can be used to produce ATP and NADPH, high-energy molecules that are necessary for growth.

Cyclic Photophosphorylation

This form of photophosphorylation occurs on the thylakoid membrane. In cyclic electron flow, the electron begins in a pigment complex called photosystem I, passes from the primary acceptor to ferredoxin, then to cytochrome b_6f (a similar complex to that found in mitochondria), and then to plastocyanin before returning to chlorophyll. This transport chain produces a proton-motive force, pumping H^+ ions across the membrane; this produces a concentration gradient that can be used to power ATP synthase during chemiosmosis. This pathway is known as cyclic photophosphorylation, and it produces neither O_2 nor NADPH. Unlike non-cyclic photophosphorylation, NADP+ does not accept the electrons; they are instead sent back to cytochrome b6f complex.

In bacterial photosynthesis, a single photosystem is used, and therefore is involved in cyclic photophosphorylation. It is favored in anaerobic conditions and conditions of high irradiance and CO_2 compensation points.

Non-cyclic Photophosphorylation

The other pathway, non-cyclic photophosphorylation, is a two-stage process involving two different chlorophyll photosystems. Being a light reaction, non-cyclic photophosphorylation occurs in the frets or stroma lamellae. First, a water molecule is broken down into $2H^+ + 1/2\ O_2 + 2e^-$ by a process called photolysis (or *water-splitting*). The two electrons from the water molecule are kept in photosystem II, while the $2H^+$ and $1/2O_2$ are left out for further use. Then a photon is absorbed by chlorophyll pigments surrounding the reaction core center of the photosystem. The light excites the electrons of each pigment, causing a chain reaction that eventually transfers energy to the core of photosystem II, exciting the two electrons that are transferred to the primary electron acceptor, pheophytin. The deficit of electrons is replenished by taking electrons from another molecule of water. The electrons transfer from pheophytin to plastoquinone, which takes the $2e^-$ from Pheophytin, and two H^+ atoms from the stroma and forms PQH_2, which later is broken into PQ, the $2e^-$ is released to Cytochrome b_6f complex and the two H^+ ions are released into thylakoid lumen. The electrons then pass through the Cyt b_6 and Cyt f. Then they are passed to plastocyanin, providing the energy for hydrogen ions (H^+) to be pumped into the thylakoid space. This creates a gradient, making H^+ ions flow back into the stroma of the chloroplast, providing the energy for the regeneration of ATP.

The photosystem II complex replaced its lost electrons from an external source; however, the two other electrons are not returned to photosystem II as they would in the analogous cyclic pathway. Instead, the still-excited electrons are transferred to a photosystem I complex, which boosts their energy level to a higher level using a second solar photon. The highly excited electrons are transferred to the acceptor molecule, but this time are passed on to an enzyme called Ferredoxin-NADP⁺ reductase which uses them to catalyse the reaction (as shown):

$$NADP^+ + 2H^+ + 2e^- \rightarrow NADPH + H^+$$

This consumes the H^+ ions produced by the splitting of water, leading to a net production of $1/2O_2$, ATP, and NADPH+H⁺ with the consumption of solar photons and water.

The concentration of NADPH in the chloroplast may help regulate which pathway electrons take through the light reactions. When the chloroplast runs low on ATP for the Calvin cycle, NADPH will accumulate and the plant may shift from noncyclic to cyclic electron flow.

Photorespiration

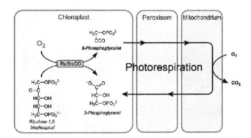

Simplified C2 Cycle

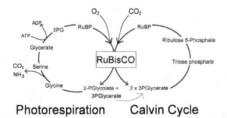

Photorespiration   Calvin Cycle

Simplified photorespiration and Calvin cycle

Photorespiration (also known as the oxidative photosynthetic carbon cycle, or C2 photosynthesis) refers to a process in plant metabolism where the enzyme RuBisCO oxygenates RuBP, causing some of the energy produced by photosynthesis to be wasted. The desired reaction is the addition of carbon dioxide to RuBP (carboxylation), a key step in the Calvin–Benson cycle, however approximately 25% of reactions by RuBisCO instead add oxygen to RuBP (oxygenation), creating a product that cannot be used within the Calvin–Benson cycle. This process reduces the efficiency of photosynthesis, potentially reducing photosynthetic output by 25% in C_3 plants. Photorespiration involves a complex network of enzyme reactions that exchange metabolites between chloroplasts, leaf peroxisomes and mitochondria.

The oxygenation reaction of RuBisCO is a wasteful process because 3-phosphoglycerate is created at a reduced rate and higher metabolic cost compared with RuBP carboxylase activity. While photorespiratory carbon cycling results in the formation of G3P eventually, there is still a net loss of carbon (around 25% of carbon fixed by photosynthesis is re-released as CO_2) and nitrogen, as ammonia. Ammonia must be detoxified at a substantial cost to the cell. Photorespiration also incurs a direct cost of one ATP and one NAD(P)H.

While it is common to refer to the entire process as photorespiration, technically the term refers only to the metabolic network which acts to rescue the products of the oxygenation reaction (phosphoglycolate).

Photorespiratory Reactions

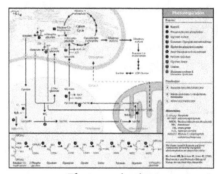

Photorespiration

Addition of molecular oxygen to ribulose-1,5-bisphosphate produces 3-phosphoglycerate (PGA) and 2-phosphoglycolate (2PG, or PG). PGA is the normal product of carboxylation, and productively enters the Calvin cycle. Phosphoglycolate, however, inhibits certain enzymes involved in photosynthetic carbon fixation (hence is often said to be an 'inhibitor of photosynthesis'). It is also relatively difficult to recycle: in higher plants it is salvaged by a series of reactions in the peroxisome, mitochondria, and again in the peroxisome where it is converted into glycerate. Glycerate reenters the chloroplast and by the same transporter that exports glycolate. A cost of 1 ATP is associated with conversion to 3-phosphoglycerate (PGA) (Phosphorylation), within the chloroplast, which is then free to re-enter the Calvin cycle.

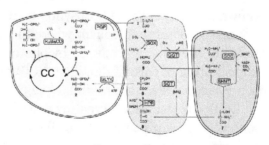

Photorespiration

There are several costs associated with this metabolic pathway; one being the production of hydrogen peroxide in the peroxisome (associated with the conversion of glycolate to glyoxylate). Hydrogen peroxide is a dangerously strong oxidant which must be immediately broken down into water and oxygen by the enzyme catalase. The conversion of 2x 2Carbon glycine to 1 C3 serine in the mitochondria by the enzyme glycine-decarboxylase is a key step, which releases CO_2, NH_3, and reduces NAD to NADH. Thus, 1 CO2 molecule is produced for every 3 molecules of O2 (two deriving from the activity of RuBisCO, the third from peroxisomal oxidations). The assimilation of NH_3 occurs via the GS-GOGAT cycle, at a cost of one ATP and one NADPH.

Cyanobacteria have three possible pathways through which they can metabolise 2-phosphoglycolate. They are unable to grow if all three pathways are knocked out, despite having a carbon concentrating mechanism that should dramatically reduce the rate of photorespiration.

Substrate Specificity of RuBisCO

The oxidative photosynthetic carbon cycle reaction is catalyzed by RuBP oxygenase activity:

RuBP + O2 → Phosphoglycolate + 3-phosphoglycerate + 2H+

Oxygenase activity of RuBisCO

During the catalysis by RuBisCO, an 'activated' intermediate is formed (an enediol intermediate) in the RuBisCO active site. This intermediate is able to react with either CO2 or O2. It has been demonstrated that the specific shape of the RuBisCO active site acts to encourage reactions with CO2. Although there is a significant "failure" rate (~25% of reactions are oxygenation rather than carboxylation), this represents significant favouring of CO2, when the relative abundance of the two gasses is taken into account: in the current atmosphere, O2 is approximately 500 times more abundant, and in solution O2 is 25X more abundant than CO2. The ability of RuBisCO to specify between the two gasses is known as its selectivity factor (or Srel), and it varies between species, with angiosperms more efficient than other plants, but with little variation among the vascular plants.

A suggested explanation into RuBisCO's inability to discriminate completely between CO2 and O2 is that it is an evolutionary relic: The early atmosphere in which primitive plants originated contained very little oxygen, the early evolution of RuBisCO was not influenced by its ability to discriminate between O2 and CO2.

Conditions Which Increase Photorespiration

Photorespiration rates are increased by:

Altered Substrate Availability: Lowered CO_2 Or increased O_2

Factors which influence this include the atmospheric abundance of the two gases, the supply of the gases to the site of fixation (i.e. in land plants: whether the stomata are open or closed), the length of the liquid phase (how far these gases have to diffuse through water in order to reach the reaction site). For example, when the stomata are closed to prevent water loss during drought: this limits the CO_2 supply, while O2 production within the leaf will continue. In algae (and plants which photosynthesise underwater); gases have to diffuse significant distances through water, which results in a decrease in the availability of CO_2 relative to O2. It has been predicted that the increase in ambient CO_2 concentrations predicted over the next 100 years may reduce the rate of photorespiration in most plants by around 50%.

Increased Temperature

At higher temperatures RuBisCO is less able to discriminate between CO_2 and O2. This is because the enediol intermediate is less stable. Increasing temperatures also reduce the solubility of CO_2, thus reducing the concentration of CO_2 relative to O2 in the chloroplast.

Biological Adaptation to Minimize Photorespiration

Certain species of plants or algae have mechanisms to reduce uptake of molecular oxygen by RuBisCO. These are commonly referred to as Carbon Concentrating Mechanisms (CCMs), as they increase the concentration of CO2 so that RuBisCO is less likely to produce glycolate through reaction with O2.

Biochemical Carbon Concentrating Mechanisms

Biochemical CCMs concentrate carbon dioxide in one temporal or spatial region, through metabolite exchange. C_4 and CAM photosynthesis both use the enzyme Phosphoenolpyruvate carboxylase (PEPC) to add CO2 to a 3-Carbon sugar. PEPC is faster than RuBisCO, and more selective for CO2.

C_4

C_4 plants capture carbon dioxide in their mesophyll cells (using an enzyme called Phosphoenolpyruvate carboxylase which catalyzes the combination of carbon dioxide with a compound called Phosphoenolpyruvate (PEP)), forming oxaloacetate. This oxaloacetate is then converted to malate and is transported into the bundle sheath cells (site of carbon dioxide fixation by RuBisCO) where oxygen concentration is low to avoid photorespiration. Here, carbon dioxide is removed from the malate and combined with RuBP by RuBisCO in the usual way, and the Calvin cycle proceeds as normal. The CO2 concentrations in the Bundle Sheath are approximately 10-20 fold higher than the concentration in the mesophyll cells.

Corn uses the C_4 pathway, minimizing photorespiration.

This ability to avoid photorespiration makes these plants more hardy than other plants in dry and hot environments, wherein stomata are closed and internal carbon dioxide levels are low. Under these conditions, photorespiration does occur in C_4 plants, but at a much reduced level compared with C_3 plants in the same conditions. C_4 plants include sugar cane, corn (maize), and sorghum.

CAM (Crassulacean Acid Metabolism)

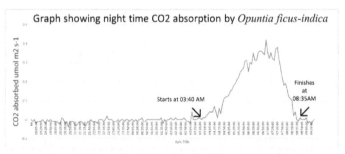

Overnight graph of CO_2 absorbed by a CAM plant

CAM plants, such as cacti and succulent plants, also use the enzyme PEP carboxylase to capture carbon dioxide, but only at night. Crassulacean acid metabolism allows plants to conduct most of their gas exchange in the cooler night-time air, sequestering carbon in 4 carbon sugars which can be released to the photosynthesizing cells during the day. This allows CAM plants to reduce water loss (transpiration) by maintaining closed stomata during the day. CAM plants usually display other water-saving characteristics, such as thick cuticles, stomata with small apertures, and typically lose around 1/3 of the amount of water per CO2 fixed.

Algae

There have been some reports of algae operating a biochemical CCM: shuttling metabolites within single cells to concentrate CO_2 in one area. This process is not fully understood.

Biophysical Carbon-concentrating Mechanisms

This type of carbon-concentrating mechanism (CCM) relies on a contained compartment within the cell into which CO_2 is shuttled, and where RuBisCO is highly expressed. In many species, biophysical CCMs are only induced under low carbon dioxide concentrations. Biophysical CCMs are more evolutionarily ancient than biochemical CCMs. There is some debate as to when biophysical CCMs first evolved, but it is likely to have been during a period of low carbon dioxide, after the Great Oxygenation Event (2.4 billion years ago). Low CO2 periods occurred around 750, 650, and 320-270 million years ago.

Eukaryotic Algae

In nearly all species of eukaryotic algae (*Chloromonas* being one notable exception), upon induction of the CCM, ~95% of RuBisCO is densely packed into a single subcellular compartment: the pyrenoid. Carbon dioxide is concentrated in this compartment using a combination of CO_2 pumps, bicarbonate pumps, and carbonic anhydrases. The pyrenoid is not a membrane bound compartment, but is found within the chloroplast, often surrounded by a starch sheath (which is not thought to serve a function in the CCM).

Hornworts

Certain species of hornwort are the only land plants which are known to have a biophysical CCM involving concentration of carbon dioxide within pyrenoids in their chloroplasts.

Cyanobacteria

Cyanobacterial CCMs are similar in principle to those found in eukaryotic algae and hornworts, but the compartment into which carbon dioxide is concentrated has several structural differences. Instead of the pyrenoid, cyanobacteria contain carboxysomes, which have a protein shell, and linker proteins packing RuBisCO inside with a very regular structure. Cyanobacterial CCMs are much better understood than those found in eukaryotes, partly due to the ease of genetic manipulation of prokaryotes.

Possible Purpose of Photorespiration

Reducing photorespiration may not result in increased growth rates for plants. Photorespiration may be necessary for the assimilation of nitrate from soil. Thus, a reduction in photorespiration by genetic engineering or because of increasing atmospheric carbon dioxide due to fossil fuel burning may not benefit plants as has been proposed. Several physiological processes may be responsible for linking photorespiration and nitrogen assimilation. Photorespiration increases availability of NADH, which is required for the conversion of nitrate to nitrite. Certain nitrite transporters also transport bicarbonate, and elevated CO_2 has been shown to suppress nitrite transport into chloroplasts.

Although photorespiration is greatly reduced in C_4 species, it is still an essential pathway - mutants without functioning 2-phosphoglycolate metabolism cannot grow in normal conditions. One mutant was shown to rapidly accumulate glycolate.

Although the functions of photorespiration remain controversial, it is widely accepted that this pathway influences a wide range of processes from bioenergetics, photosystem II function, and carbon metabolism to nitrogen assimilation and respiration. The photorespiratory pathway is a major source of hydrogen peroxide (H2O2) in photosynthetic cells. Through H2O2 production and pyridine nucleotide interactions, photorespiration makes a key contribution to cellular redox homeostasis. In so doing, it influences multiple signalling pathways, in particular, those that govern plant hormonal responses controlling growth, environmental and defense responses, and programmed cell death.

Another theory postulates that it may function as a "safety valve", preventing the excess of reductive potential coming from an overreduced NADPH-pool from reacting with oxygen and producing free radicals, as these can damage the metabolic functions of the cell by subsequent oxidation of membrane lipids, proteins or nucleotides.

Photosynthate Partitioning

Photosynthate partitioning is the deferential distribution of photosynthates to plant tissues. A photosynthate is the resulting product of photosynthesis, these products are generally sugars. These sugars that are created from photosynthesis are broken down to create energy for use by the plant. Sugar and other compounds move via the phloem to tissues that have an energy demand. These areas of demand are called sinks. While areas with an excess of sugars and a low energy demand are called sources. Many times sinks are the actively growing tissues of the plant while the sources are where sugars are produced by photosynthesis—the leaves of plants. Sugars are actively loaded into the phloem and moved by a positive pressure flow created by solute concentrations and turgor pressure between xylem and phloem vessel elements (specialized plant cells). This movement of sugars is referred to as translocation. When sugars arrive at the sink they are unloaded for storage or broken down/metabolized.

The partitioning of these sugars depends on multiple factors such as the vascular connections that exist, the location of the sink to source, the developmental stage, and the strength of that sink. Vascular connections exist between sources and sinks and those that are the most direct have been shown to receive more photosynthates than those that must travel through extensive connections. This also goes for proximity those closer to the source are easier to translocate sugars to. Developmental stage plays a large role in partitioning, organs that are young such as meristems and new leaves have a higher demand, as well as those that are entering reproductive maturity—creating fruits, flowers, and seeds. Many of these developing organs have a higher sink strength. Those with higher sink strengths receive more photosynthates than lower strength sinks. Sinks compete to receive these compounds and combination of factors playing in determining how much and how fast sinks receives photosynthates to grow and complete physiological activities.

Photosynthetically Active Radiation

Photosynthetically active radiation, often abbreviated PAR, designates the spectral range (wave band) of solar radiation from 400 to 700 nanometers that photosynthetic organisms are able to use in the process of photosynthesis. This spectral region corresponds more or less with the range of light visible to the human eye. Photons at shorter wavelengths tend to be so energetic that they can be damaging to cells and tissues, but are mostly filtered out by the ozone layer in the stratosphere. Photons at longer wavelengths do not carry enough energy to allow photosynthesis to take place.

Other living organisms, such as Cyanobacteria, purple bacteria and Heliobacteria, can exploit solar light in slightly extended spectral regions, such as the near-infrared. These bacteria live in environments such as the bottom of stagnant ponds, sediment and ocean depths. Because of their pigments, they form colorful mats of green, red and purple.

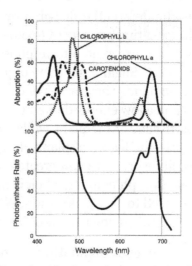

Typical PAR action spectrum, shown beside absorption spectra for chlorophyll-A, chlorophyll-B, and carotenoids

Chlorophyll, the most abundant plant pigment, is most efficient in capturing red and blue light. Accessory pigments such as carotenes and xanthophylls harvest some green light and pass it on to the photosynthetic process, but enough of the green wavelengths are reflected to give leaves their characteristic color. An exception to the predominance of chlorophyll is autumn, when chlorophyll is degraded (because it contains N and Mg) but the accessory pigments are not (because they only contain C, H and O) and remain in the leaf producing red, yellow and orange leaves.

PAR measurement is used in agriculture, forestry and oceanography. One of the requirements for productive farmland is adequate PAR, so PAR is used to evaluate agricultural investment potential. PAR sensors stationed at various levels of the forest canopy measure the pattern of PAR availability and utilization. Photosynthetic rate and related parameters can be measured non-destructively using a photosynthesis system, and these instruments measure PAR and sometimes control PAR at set intensities. PAR measurements are also used to calculate the euphotic depth in the ocean.

Units

The irradiance of PAR can be measured in energy units (W/m^2), which is relevant in energy-balance considerations for photosynthetic organisms.

However, photosynthesis is a quantum process and the chemical reactions of photosynthesis are more dependent on the number of photons than the energy contained in the photons. Therefore plant biologists often quantify PAR using the number of photons in the 400-700 nm range received by a surface for a specified amount of time, or the Photosynthetic Photon Flux Density (PPFD). This is normally measured using mol $m^{-2}s^{-1}$. PPFD used to sometimes be expressed as einstein units, i.e., $\mu E\ m^{-2}s^{-1}$, although this usage is nonstandard and no longer used.

The conversion between energy-based PAR and photon-based PAR depends on the spectrum of the light source. The following table shows the conversion factors from watts for black-body spectra that are truncated to the range 400–700 nm. It also shows the luminous efficacy for these light sources and the fraction of a real black-body radiator that is emitted as PAR.

T (K)	η_v (lm/W*)	η_photon (μmol/J* or μmol s⁻¹W*⁻¹)	η_photon (mol day⁻¹ W*⁻¹)	η_PAR (W*/W)
3000 (warm white)	269	4.98	0.43	0.0809
4000	277	4.78	0.413	0.208
5800 (daylight)	265	4.56	0.394	0.368
Note: W* and J* indicates PAR watts and PAR joules (400–700 nm).				

For example, a light source of 1000 lm at a color temperature of 5800 K would emit approximately 1000/265 = 3.8 W of PAR, which is equivalent to 3.8*4.56 = 17.3 µmol/s. For a black-body light source at 5800 K, such as the sun is approximately, a fraction 0.368 of its total emitted radiation is emitted as PAR. For artificial light sources, that usually do not have a black-body spectrum, these conversion factors are only approximate.

The quantities in the table are calculated as

$$\eta_v(T) = \frac{\int_{\lambda_1}^{\lambda_2} B(\lambda, T) 683 \, [\text{lm/W}] y(\lambda) d\lambda}{\int_{\lambda_1}^{\lambda_2} B(\lambda, T) d\lambda},$$

$$\eta_{\text{photon}}(T) = \frac{\int_{\lambda_1}^{\lambda_2} B(\lambda, T) \frac{\lambda}{hcN_A} d\lambda}{\int_{\lambda_1}^{\lambda_2} B(\lambda, T) d\lambda},$$

$$\eta_{\text{PAR}}(T) = \frac{\int_{\lambda_1}^{\lambda_2} B(\lambda, T) d\lambda}{\int_{0}^{\infty} B(\lambda, T) d\lambda},$$

where $B(\lambda, T)$ is the black-body spectrum according to Planck's law, y is the standard luminosity function, λ_1, λ_2 represent the wavelength range (400 700 nm) of PAR, and N_A is the Avogadro constant.

Yield Photon Flux

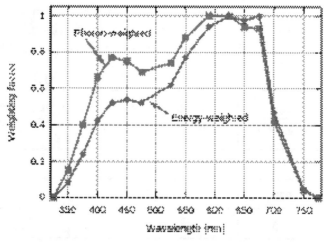

Weighting factor for photosynthesis. The photon-weighted curve is for converting PPFD to YPF; the energy-weighted curve is for weighting PAR expressed in watts or joules.

PAR as described above does not distinguish between different wavelengths between 400 and 700 nm, and assumes that wavelengths outside this range have zero photosynthetic action. If the exact spectrum of the light is known, the photosynthetic photon flux density (PPFD) values in μmol/s can be modified by applying different weighting factor to different wavelengths. This results in a quantity called the yield photon flux (YPF). The red curve in the graph shows that photons around 610 nm (orange-red) have the highest amount of photosynthesis per photon. However, because short-wavelength photons carry more energy per photon, the maximum amount of photosynthesis per incident unit of energy is at a longer wavelength, around 650 nm (deep red).

It has been noted that there is considerable misunderstanding over the effect of light quality on plant growth and many manufacturers claim significantly increased plant growth due to light quality (spectral distribution or the ratio of the colors). A widely used estimate of the effect of light quality on photosynthesis comes from the Yield Photon Flux (YPF) curve, which indicates that orange and red photons between 600 to 630 nm can result in 20 to 30% more photosynthesis than blue or cyan photons between 400 and 540 nm.

The YPF curve was developed from short-term measurements made on single leaves in low light. Some longer-term studies with whole plants in higher light indicate that light quality may have a smaller effect on plant growth rate than light quantity.

Chloroplast

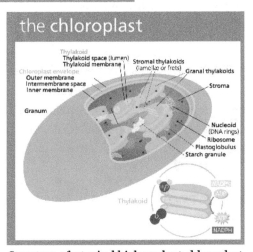

Structure of a typical higher-plant chloroplast

Chloroplasts are organelles, specialized subunits, in plant and algal cells. Their discovery inside plant cells is usually credited to Julius von Sachs (1832–1897), an influential botanist and author of standard botanical textbooks – sometimes called "The Father of Plant Physiology".

Chloroplasts' main role is to conduct photosynthesis, where the photosynthetic pigment chlorophyll captures the energy from sunlight and converts it and stores it in the energy-storage molecules ATP and NADPH while freeing oxygen from water. They then use the ATP and NADPH to make organic molecules from carbon dioxide in a process known as the Calvin cycle. Chloroplasts carry out a number of other functions, including fatty acid synthesis, much amino acid synthesis, and the immune response in plants. The number of chloroplasts per cell varies from 1 in algae up to 100 in plants like Arabidopsis and wheat.

A chloroplast is one of three types of plastids, characterized by its high concentration of chlorophyll, the other two types, the leucoplast and the chromoplast, contain little chlorophyll and do not carry out photosynthesis.

Chloroplasts are highly dynamic—they circulate and are moved around within plant cells, and occasionally pinch in two to reproduce. Their behavior is strongly influenced by environmental factors like light color and intensity. Chloroplasts, like mitochondria, contain their own DNA, which is thought to be inherited from their ancestor—a photosynthetic cyanobacterium that was engulfed by an early eukaryotic cell. Chloroplasts cannot be made by the plant cell and must be inherited by each daughter cell during cell division.

With one exception (the amoeboid *Paulinella chromatophora*), all chloroplasts can probably be traced back to a single endosymbiotic event, when a cyanobacterium was engulfed by the eukaryote. Despite this, chloroplasts can be found in an extremely wide set of organisms, some not even directly related to each other—a consequence of many secondary and even tertiary endosymbiotic events.

Discovery

The first definitive description of a chloroplast (*Chlorophyllkörnen*, "grain of chlorophyll") was given by Hugo von Mohl in 1837 as discrete bodies within the green plant cell. In 1883, A. F. W. Schimper would name these bodies as "chloroplastids" (*Chloroplastiden*). In 1884, Eduard Strasburger adopted the term "chloroplasts" (*Chloroplasten*).

Chloroplast Lineages and Evolution

Chloroplasts are one of many types of organelles in the plant cell. They are considered to have originated from cyanobacteria through endosymbiosis—when a eukaryotic cell engulfed a photosynthesizing cyanobacterium that became a permanent resident in the cell. Mitochondria are thought to have come from a similar event, where an aerobic prokaryote was engulfed. This origin of chloroplasts was first suggested by the Russian biologist Konstantin Mereschkowski in 1905 after Andreas Schimper observed in 1883 that chloroplasts closely resemble cyanobacteria. Chloroplasts are only found in plants and algae.

Cyanobacterial Ancestor

Cyanobacteria are considered the ancestors of chloroplasts. They are sometimes called blue-green algae even though they are prokaryotes. They are a diverse phylum of bacteria capable of carrying out photosynthesis, and are gram-negative, meaning that they have two cell membranes. Cyanobacteria also contain a peptidoglycan cell wall, which is thicker than in other gram-negative bacteria, and which is located between their two cell membranes. Like chloroplasts, they have thylakoids within. On the thylakoid membranes are photosynthetic pigments, including chlorophyll a. Phycobilins are also common cyanobacterial pigments, usually organized into hemispherical phycobilisomes attached to the outside of the thylakoid membranes (phycobilins are not shared with all chloroplasts though).

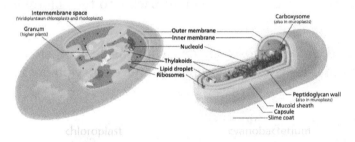

Both chloroplasts and cyanobacteria have a double membrane, DNA, ribosomes, and thylakoids. Both the chloroplast and cyanobacterium depicted are idealized versions (the chloroplast is that of a higher plant)—a lot of diversity exists among chloroplasts and cyanobacteria.

Primary Endosymbiosis

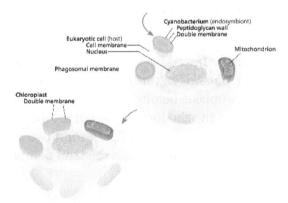

Primary Endosymbiosis

A eukaryote with mitochondria engulfed a cyanobacterium in an event of serial primary endosymbiosis, creating a lineage of cells with both organelles. It is important to note that the cyanobacterial endosymbiont already had a double membrane—the phagosomal vacuole-derived membrane was lost.

Somewhere around a billion years ago, a free-living cyanobacterium entered an early eukaryotic cell, either as food or as an internal parasite, but managed to escape the phagocytic vacuole it was contained in. The two innermost lipid-bilayer membranes that surround all chloroplasts correspond to the outer and inner membranes of the ancestral cyanobacterium's gram negative cell wall, and not the phagosomal membrane from the host, which was probably lost. The new cellular resident quickly became an advantage, providing food for the eukaryotic host, which allowed it to live within it. Over time, the cyanobacterium was assimilated, and many of its genes were lost or transferred to the nucleus of the host. Some of its proteins were then synthesized in the cytoplasm of the host cell, and imported back into the chloroplast (formerly the cyanobacterium).

This event is called *endosymbiosis*, or "cell living inside another cell". The cell living inside the other cell is called the *endosymbiont*; the endosymbiont is found inside the *host cell*.

Chloroplasts are believed to have arisen after mitochondria, since all eukaryotes contain mitochondria, but not all have chloroplasts. This is called *serial endosymbiosis*— an early eukaryote engulfing the mitochondrion ancestor, and some descendants of it then engulfing the chloroplast ancestor, creating a cell with both chloroplasts and mitochondria.

Whether or not chloroplasts came from a single endosymbiotic event, or many independent engulfments across various eukaryotic lineages, has been long debated, but it is now generally held that all organisms with chloroplasts either share a single ancestor or obtained their chloroplast from organisms that share a common ancestor that took in a cyanobacterium 600–1600 million years ago.

These chloroplasts, which can be traced back directly to a cyanobacterial ancestor, are known as *primary plastids* (*"plastid"* in this context means almost the same thing as chloroplast). All primary chloroplasts belong to one of three chloroplast lineages—the glaucophyte chloroplast lineage, the rhodophyte, or red algal chloroplast lineage, or the chloroplastidan, or green chloroplast lineage. The second two are the largest, and the green chloroplast lineage is the one that contains the land plants.

Glaucophyta

The alga *Cyanophora*, a glaucophyte, is thought to be one of the first organisms to contain a chloroplast. The glaucophyte chloroplast group is the smallest of the three primary chloroplast lineages, being found in only 13 species, and is thought to be the one that branched off the earliest. Glaucophytes have chloroplasts that retain a peptidoglycan wall between their double membranes, like their cyanobacterial parent. For this reason, glaucophyte chloroplasts are also known as *muroplasts*. Glaucophyte chloroplasts also contain concentric unstacked thylakoids, which surround a carboxysome - an icosahedral structure that glaucophyte chloroplasts and cyanobacteria keep their carbon fixation enzyme rubisco in. The starch that they synthesize collects outside the chloroplast. Like cyanobacteria, glaucophyte chloroplast thylakoids are studded with light collecting structures called phycobilisomes. For these reasons, glaucophyte chloroplasts are considered a primitive intermediate between cyanobacteria and the more evolved chloroplasts in red algae and plants.

Diversity of red algae Clockwise from top left: *Bornetia secundiflora*, *Peyssonnelia squamaria*, *Cyanidium*, *Laurencia*, *Callophyllis laciniata*. Red algal chloroplasts are characterized by phycobilin pigments which often give them their reddish color.

Rhodophyceae (Red Algae)

The rhodophyte, or red algal chloroplast group is another large and diverse chloroplast lineage. Rhodophyte chloroplasts are also called *rhodoplasts*, literally "red chloroplasts".

Rhodoplasts have a double membrane with an intermembrane space and phycobilin pigments organized into phycobilisomes on the thylakoid membranes, preventing their thylakoids from stacking. Some contain pyrenoids. Rhodoplasts have chlorophyll *a* and phycobilins for photosynthetic pigments; the phycobilin phycoerytherin is responsible for giving many red algae their distinctive red color. However, since they also contain the blue-green chlorophyll *a* and other pigments, many are reddish to purple from the combination. The red phycoerytherin pigment is an adaptation to help red algae catch more sunlight in deep water—as such, some red algae that live in shallow water have less phycoerytherin in their rhodoplasts, and can appear more greenish. Rhodoplasts synthesize a form of starch called floridean starch, which collects into granules outside the rhodoplast, in the cytoplasm of the red alga.

Chloroplastida (Green Algae and Plants)

Diversity of green algae Clockwise from top left: *Scenedesmus*, *Micrasterias*, *Hydrodictyon*, *Stigeoclonium*, *Volvox*. Green algal chloroplasts are characterized by their pigments chlorophyll *a* and chlorophyll *b* which give them their green color.

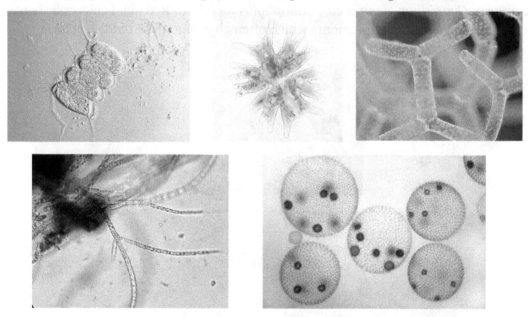

The chloroplastidan chloroplasts, or green chloroplasts, are another large, highly diverse primary chloroplast lineage. Their host organisms are commonly known as the green algae and land plants. They differ from glaucophyte and red algal chloroplasts in that they have lost their phycobilisomes, and contain chlorophyll *b* instead. Most green chloroplasts are (obviously) green, though some aren't, like some forms of *Hæmatococcus pluvialis*, due to accessory pigments that override the chlorophylls' green colors. Chloroplastidan chloroplasts have lost the peptidoglycan wall between their double membrane, and have replaced it with an intermembrane space. Some plants seem to have kept the genes for the synthesis of the peptidoglycan layer, though they've been repurposed for use in chloroplast division instead.

Green algae and plants keep their starch *inside* their chloroplasts, and in plants and some algae, the chloroplast thylakoids are arranged in grana stacks. Some green algal chloroplasts contain a structure called a pyrenoid, which is functionally similar to the glaucophyte carboxysome in that it is where rubisco and CO_2 are concentrated in the chloroplast.

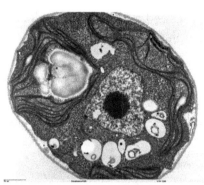

Transmission electron micrograph of Chlamydomonas reinhardtii, a green alga that contains a pyrenoid surrounded by starch.

Helicosporidium

Helicosporidium is a genus of nonphotosynthetic parasitic green algae that is thought to contain a vestigial chloroplast. Genes from a chloroplast and nuclear genes indicating the presence of a chloroplast have been found in Helicosporidium even if nobody's seen the chloroplast itself.

Secondary and Tertiary Endosymbiosis

Many other organisms obtained chloroplasts from the primary chloroplast lineages through secondary endosymbiosis—engulfing a red or green alga that contained a chloroplast. These chloroplasts are known as secondary plastids.

While primary chloroplasts have a double membrane from their cyanobacterial ancestor, secondary chloroplasts have additional membranes outside of the original two, as a result of the secondary endosymbiotic event, when a nonphotosynthetic eukaryote engulfed a chloroplast-containing alga but failed to digest it—much like the cyanobacterium at the beginning of this story. The engulfed alga was broken down, leaving only its chloroplast, and sometimes its cell membrane and nucleus, forming a chloroplast with three or four membranes—the two cyanobacterial membranes, sometimes the eaten alga's cell membrane, and the phagosomal vacuole from the host's cell membrane.

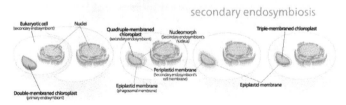

Secondary endosymbiosis consisted of a eukaryotic alga being engulfed by another eukaryote, forming a chloroplast with three or four membranes.

The genes in the phagocytosed eukaryote's nucleus are often transferred to the secondary host's nucleus. Cryptomonads and chlorarachniophytes retain the phagocytosed eukaryote's nucleus, an object called a nucleomorph, located between the second and third membranes of the chloroplast.

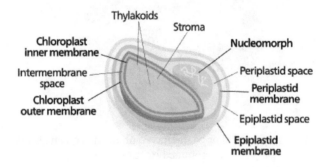

Diagram of a four membraned chloroplast containing a nucleomorph.

All secondary chloroplasts come from green and red algae—no secondary chloroplasts from glaucophytes have been observed, probably because glaucophytes are relatively rare in nature, making them less likely to have been taken up by another eukaryote.

Green algal Derived Chloroplasts

Green algae have been taken up by the euglenids, chlorarachniophytes, a lineage of dinoflagellates, and possibly the ancestor of the chromalveolates in three or four separate engulfments. Many green algal derived chloroplasts contain pyrenoids, but unlike chloroplasts in their green algal ancestors, starch collects in granules outside the chloroplast.

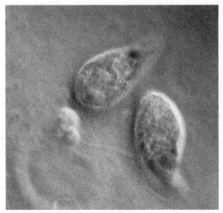

Euglena, a euglenophyte, contains secondary chloroplasts from green algae.

Euglenophytes

Euglenophytes are a group of common flagellated protists that contain chloroplasts derived from a green alga. Euglenophyte chloroplasts have three membranes—it is thought that the membrane of the primary endosymbiont was lost, leaving the cyanobacterial membranes, and the secondary host's phagosomal membrane. Euglenophyte chloroplasts have a pyrenoid and thylakoids stacked in groups of three. Starch is stored in the form of paramylon, which is contained in membrane-bound granules in the cytoplasm of the euglenophyte.

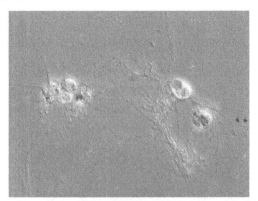

Chlorarachnion reptans is a chlorarachniophyte. Chlorarachniophytes replaced their original red algal endosymbiont with a green alga.

Chlorarachniophytes

Chlorarachniophytes are a rare group of organisms that also contain chloroplasts derived from green algae, though their story is more complicated than that of the euglenophytes. The ancestor of chlorarachniophytes is thought to have been a chromalveolate, a eukaryote with a *red* algal derived chloroplast. It is then thought to have lost its first red algal chloroplast, and later engulfed a green alga, giving it its second, green algal derived chloroplast.

Chlorarachniophyte chloroplasts are bounded by four membranes, except near the cell membrane, where the chloroplast membranes fuse into a double membrane. Their thylakoids are arranged in loose stacks of three. Chlorarachniophytes have a form of starch called chrysolaminarin, which they store in the cytoplasm, often collected around the chloroplast pyrenoid, which bulges into the cytoplasm.

Chlorarachniophyte chloroplasts are notable because the green alga they are derived from has not been completely broken down—its nucleus still persists as a nucleomorph found between the second and third chloroplast membranes—the periplastid space, which corresponds to the green alga's cytoplasm.

Early Chromalveolates

Recent research has suggested that the ancestor of the chromalveolates acquired a green algal prasinophyte endosymbiont. The green algal derived chloroplast was lost and replaced with a red algal derived chloroplast, but not before contributing some of its genes to the early chromalveolate's nucleus. The presence of both green algal and red algal genes in chromalveolates probably helps them thrive under fluctuating light conditions.

Red Algal Derived Chloroplasts (Chromalveolate Chloroplasts)

Like green algae, red algae have also been taken up in secondary endosymbiosis, though it is thought that all red algal derived chloroplasts are descended from a single red alga that was engulfed by an early chromalveolate, giving rise to the chromalveolates, some of which, like the ciliates, subsequently lost the chloroplast. This is still debated though.

Pyrenoids and stacked thylakoids are common in chromalveolate chloroplasts, and the outermost membrane of many are continuous with the rough endoplasmic reticulum and studded with ribosomes. They have lost their phycobilisomes and exchanged them for chlorophyll *c*, which isn't found in primary red algal chloroplasts themselves.

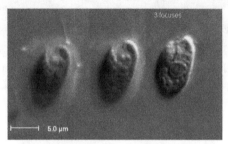

Rhodomonas salina is a cryptophyte.

Cryptophytes

Cryptophytes, or cryptomonads are a group of algae that contain a red-algal derived chloroplast. Cryptophyte chloroplasts contain a nucleomorph that superficially resembles that of the chlorarachniophytes. Cryptophyte chloroplasts have four membranes, the outermost of which is continuous with the rough endoplasmic reticulum. They synthesize ordinary starch, which is stored in granules found in the periplastid space—outside the original double membrane, in the place that corresponds to the red alga's cytoplasm. Inside cryptophyte chloroplasts is a pyrenoid and thylakoids in stacks of two.

Their chloroplasts do not have phycobilisomes, but they do have phycobilin pigments which they keep in their thylakoid space, rather than anchored on the outside of their thylakoid membranes.

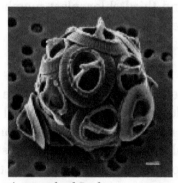

Scanning electron micrograph of Gephyrocapsa oceanica, a haptophyte.

Haptophytes

Haptophytes are similar and closely related to cryptophytes, and are thought to be the first chromalveolates to branch off. Their chloroplasts lack a nucleomorph, their thylakoids are in stacks of three, and they synthesize chrysolaminarin sugar, which they store completely outside of the chloroplast, in the cytoplasm of the haptophyte.

Heterokontophytes (Stramenopiles)

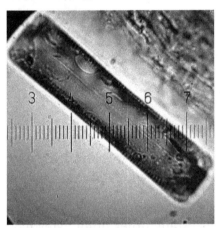

The photosynthetic pigments present in their chloroplasts give diatoms a greenish-brown color.

The heterokontophytes, also known as the stramenopiles, are a very large and diverse group of algae that also contain red algal derived chloroplasts. Heterokonts include the diatoms and the brown algae, golden algae, and yellow-green algae.

Heterokont chloroplasts are very similar to haptophyte chloroplasts, containing a pyrenoid, triplet thylakoids, and with some exceptions, having an epiplastid membrane connected to the endoplasmic reticulum. Like haptophytes, heterokontophytes store sugar in chrysolaminarin granules in the cytoplasm. Heterokontophyte chloroplasts contain chlorophyll *a* and with a few exceptions chlorophyll *c*, but also have carotenoids which give them their many colors.

Apicomplexans

Apicomplexans are another group of chromalveolates. Like the helicosproidia, they're parasitic, and have a nonphotosynthetic chloroplast. They were once thought to be related to the helicosproidia, but it is now known that the helicosproida are green algae rather than chromalveolates. The apicomplexans include *Plasmodium*, the malaria parasite. Many apicomplexans keep a vestigial red algal derived chloroplast called an apicoplast, which they inherited from their ancestors. Other apicomplexans like *Cryptosporidium* have lost the chloroplast completely. Apicomplexans store their energy in amylopectin starch granules that are located in their cytoplasm, even though they are nonphotosynthetic.

Apicoplasts have lost all photosynthetic function, and contain no photosynthetic pigments or true thylakoids. They are bounded by four membranes, but the membranes are not connected to the endoplasmic reticulum. The fact that apicomplexans still keep their nonphotosynthetic chloroplast around demonstrates how the chloroplast carries out important functions other than photosynthesis. Plant chloroplasts provide plant cells with many important things besides sugar, and apicoplasts are no different—they synthesize fatty acids, isopentenyl pyrophosphate, iron-sulfur clusters, and carry out part of the heme pathway. This makes the apicoplast an attractive target for drugs to cure apicomplexan-related diseases. The most important apicoplast function is isopentenyl pyrophosphate synthesis—in fact, apicomplexans die when something interferes with this apicoplast function, and when apicomplexans are grown in an isopentenyl pyrophosphate-rich medium, they dump the organelle.

Dinophytes

The dinoflagellates are yet another very large and diverse group of protists, around half of which are (at least partially) photosynthetic.

Most dinophyte chloroplasts are secondary red algal derived chloroplasts, like other chromalveolate chloroplasts. Many other dinophytes have lost the chloroplast (becoming the nonphotosynthetic kind of dinoflagellate), or replaced it though *tertiary* endosymbiosis—the engulfment of another chromalveolate containing a red algal derived chloroplast. Others replaced their original chloroplast with a green algal derived one.

Most dinophyte chloroplasts contain at least the photosynthetic pigments chlorophyll *a*, chlorophyll c_2, *beta*-carotene, and at least one dinophyte-unique xanthophyll (peridinin, dinoxanthin, or diadinoxanthin), giving many a golden-brown color. All dinophytes store starch in their cytoplasm, and most have chloroplasts with thylakoids arranged in stacks of three.

Peridinin-containing Dinophyte Chloroplast

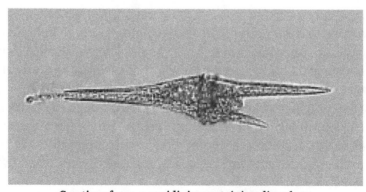

Ceratium furca, a peridinin-containing dinophyte

The most common dinophyte chloroplast is the peridinin-type chloroplast, character-ized by the carotenoid pigment peridinin in their chloroplasts, along with chlorophyll a and chlorophyll c_2. Peridinin is not found in any other group of chloroplasts. The peridinin chloroplast is bounded by three membranes (occasionally two), having lost the red algal endosymbiont's original cell membrane. The outermost membrane is not connected to the endoplasmic reticulum. They contain a pyrenoid, and have trip-let-stacked thylakoids. Starch is found outside the chloroplast. An important feature of these chloroplasts is that their chloroplast DNA is highly reduced and fragmented into many small circles. Most of the genome has migrated to the nucleus, and only critical photosynthesis-related genes remain in the chloroplast.

The peridinin chloroplast is thought to be the dinophytes' "original" chloroplast, which has been lost, reduced, replaced, or has company in several other dinophyte lineages.

Fucoxanthin-containing Dinophyte Chloroplasts (Haptophyte Endosymbionts)

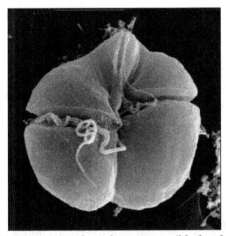

Karenia brevis is a fucoxanthin-containing dynophyte responsible for algal blooms called "red tides".

The fucoxanthin dinophyte lineages (including *Karlodinium* and *Karenia*) lost their original red algal derived chloroplast, and replaced it with a new chloroplast derived from a haptophyte endosymbiont. *Karlodinium* and *Karenia* probably took up differ-ent heterokontophytes. Because the haptophyte chloroplast has four membranes, ter-tiary endosymbiosis would be expected to create a six membraned chloroplast, adding the haptophyte's cell membrane and the dinophyte's phagosomal vacuole. However, the haptophyte was heavily reduced, stripped of a few membranes and its nucleus, leav-ing only its chloroplast (with its original double membrane), and possibly one or two additional membranes around it.

Fucoxanthin-containing chloroplasts are characterized by having the pigment fucoxan-thin (actually 19'-hexanoyloxy-fucoxanthin and/or 19'-butanoyloxy-fucoxanthin) and no peridinin. Fucoxanthin is also found in haptophyte chloroplasts, providing evidence of ancestry.

Dinophysis acuminata has chloroplasts taken from a cryptophyte.

Cryptophyte Derived Dinophyte Chloroplast

Members of the genus *Dinophysis* have a phycobilin-containing chloroplast taken from a cryptophyte. However, the cryptophyte is not an endosymbiont—only the chloroplast seems to have been taken, and the chloroplast has been stripped of its nucleomorph and outermost two membranes, leaving just a two-membraned chloroplast. Cryptophyte chloroplasts require their nucleomorph to maintain themselves, and *Dinophysis* species grown in cell culture alone cannot survive, so it is possible (but not confirmed) that the *Dinophysis* chloroplast is a kleptoplast—if so, *Dinophysis* chloroplasts wear out and *Dinophysis* species must continually engulf cryptophytes to obtain new chloroplasts to replace the old ones.

Diatom Derived Dinophyte Chloroplasts

Some dinophytes, like *Kryptoperidinium* and *Durinskia* have a diatom (heterokontophyte) derived chloroplast. These chloroplasts are bounded by up to *five* membranes, (depending on whether you count the entire diatom endosymbiont as the chloroplast, or just the red algal derived chloroplast inside it). The diatom endosymbiont has been reduced relatively little—it still retains its original mitochondria, and has endoplasmic reticulum, ribosomes, a nucleus, and of course, red algal derived chloroplasts—practically a complete cell, all inside the host's endoplasmic reticulum lumen. However the diatom endosymbiont can't store its own food—its starch is found in granules in the dinophyte host's cytoplasm instead. The diatom endosymbiont's nucleus is present, but it probably can't be called a nucleomorph because it shows no sign of genome reduction, and might have even been *expanded*. Diatoms have been engulfed by dinoflagellates at least three times.

The diatom endosymbiont is bounded by a single membrane, inside it are chloroplasts with four membranes. Like the diatom endosymbiont's diatom ancestor, the chloroplasts have triplet thylakoids and pyrenoids.

In some of these genera, the diatom endosymbiont's chloroplasts aren't the only chloroplasts in the dinophyte. The original three-membraned peridinin chloroplast is still around, converted to an eyespot.

Prasinophyte (Green Algal) Derived Dinophyte Chloroplast

Lepidodinium viride and its close relatives are dinophytes that lost their original peridinin chloroplast and replaced it with a green algal derived chloroplast (more specifically, a prasinophyte). *Lepidodinium* is the only dinophyte that has a chloroplast that's not from the rhodoplast lineage. The chloroplast is surrounded by two membranes and has no nucleomorph—all the nucleomorph genes have been transferred to the dinophyte nucleus. The endosymbiotic event that led to this chloroplast was serial secondary endosymbiosis rather than tertiary endosymbiosis—the endosymbiont was a green alga containing a primary chloroplast (making a secondary chloroplast).

Chromatophores

While most chloroplasts originate from that first set of endosymbiotic events, *Paulinella chromatophora* is an exception that acquired a photosynthetic cyanobacterial endosymbiont more recently. It is not clear whether that symbiont is closely related to the ancestral chloroplast of other eukaryotes. Being in the early stages of endosymbiosis, *Paulinella chromatophora* can offer some insights into how chloroplasts evolved. *Paulinella* cells contain one or two sausage shaped blue-green photosynthesizing structures called chromatophores, descended from the cyanobacterium *Synechococcus*. Chromatophores cannot survive outside their host. Chromatophore DNA is about a million base pairs long, containing around 850 protein encoding genes—far less than the three million base pair *Synechococcus* genome, but much larger than the approximately 150,000 base pair genome of the more assimilated chloroplast. Chromatophores have transferred much less of their DNA to the nucleus of their host. About 0.3–0.8% of the nuclear DNA in *Paulinella* is from the chromatophore, compared with 11–14% from the chloroplast in plants.

Kleptoplastidy

In some groups of mixotrophic protists, like some dinoflagellates, chloroplasts are separated from a captured alga or diatom and used temporarily. These klepto chloroplasts may only have a lifetime of a few days and are then replaced.

Chloroplast DNA

Chloroplasts have their own DNA, often abbreviated as ctDNA, or cpDNA. It is also known as the plastome. Its existence was first proved in 1962, and first sequenced in 1986—when two Japanese research teams sequenced the chloroplast DNA of liverwort and tobacco. Since then, hundreds of chloroplast DNAs from various species have been sequenced, but they're mostly those of land plants and green algae—glaucophytes, red algae, and other algal groups are extremely underrepresented, potentially introducing some bias in views of "typical" chloroplast DNA structure and content.

Molecular Structure

Chloroplast DNA Interactive gene map of chloroplast DNA from *Nicotiana tabacum*. Segments with labels on the inside reside on the B strand of DNA, segments with labels on the outside are on the A strand. Notches indicate introns.

With few exceptions, most chloroplasts have their entire chloroplast genome combined into a single large circular DNA molecule, typically 120,000–170,000 base pairs long. They can have a contour length of around 30–60 micrometers, and have a mass of about 80–130 million daltons.

While usually thought of as a circular molecule, there is some evidence that chloroplast DNA molecules more often take on a linear shape.

Inverted Repeats

Many chloroplast DNAs contain two *inverted repeats*, which separate a long single copy section (LSC) from a short single copy section (SSC). While a given pair of inverted repeats are rarely completely identical, they are always very similar to each other, apparently resulting from concerted evolution.

The inverted repeats vary wildly in length, ranging from 4,000 to 25,000 base pairs long each and containing as few as four or as many as over 150 genes. Inverted repeats in plants tend to be at the upper end of this range, each being 20,000–25,000 base pairs long.

The inverted repeat regions are highly conserved among land plants, and accumulate few mutations. Similar inverted repeats exist in the genomes of cyanobacteria and the other two chloroplast lineages (glaucophyta and rhodophyceae), suggesting that they predate the chloroplast, though some chloroplast DNAs have since lost or flipped the inverted repeats (making them direct repeats). It is possible that the inverted repeats help stabilize the rest of the chloroplast genome, as chloroplast DNAs which have lost some of the inverted repeat segments tend to get rearranged more.

Nucleoids

New chloroplasts may contain up to 100 copies of their DNA, though the number of chloroplast DNA copies decreases to about 15–20 as the chloroplasts age. They are usually packed into nucleoids, which can contain several identical chloroplast DNA rings. Many nucleoids can be found in each chloroplast. In primitive red algae, the chloroplast DNA nucleoids are clustered in the center of the chloroplast, while in green plants and green algae, the nucleoids are dispersed throughout the stroma.

Though chloroplast DNA is not associated with true histones, in red algae, similar proteins that tightly pack each chloroplast DNA ring into a nucleoid have been found.

DNA Replication

The Leading Model of cpDNA Replication

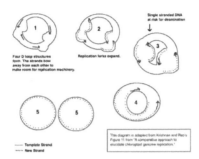

Chloroplast DNA replication via multiple D loop mechanisms. Adapted from Krishnan NM, Rao BJ's paper "A comparative approach to elucidate chloroplast genome replication."

The mechanism for chloroplast DNA (cpDNA) replication has not been conclusively determined, but two main models have been proposed. Scientists have attempted to observe chloroplast replication via electron microscopy since the 1970s. The results of the microscopy experiments led to the idea that chloroplast DNA replicates using a double displacement loop (D-loop). As the D-loop moves through the circular DNA, it adopts a theta intermediary form, also known as a Cairns replication intermediate, and completes replication with a rolling circle mechanism. Transcription starts at specific points of origin. Multiple replication forks open up, allowing replication machinery to transcribe the DNA. As replication continues, the forks grow and eventually converge. The new cpDNA structures separate, creating daughter cpDNA chromosomes.

In addition to the early microscopy experiments, this model is also supported by the amounts of deamination seen in cpDNA. Deamination occurs when an amino group is lost and is a mutation that often results in base changes. When adenine is deaminated, it becomes hypoxanthine. Hypoxanthine can bind to cytosine, and when the XC base pair is replicated, it becomes a GC (thus, an A → G base change).

Original DNA Strand
...CCATGCATGGATC...

Deamination of an Adenine
...CCATGCATGGATC...
↓
...CCHTGCATGGATC...

During Replication, H pairs with C
...CCHTGCATGGATC...
...GGCACGTACCTAG...

When Replicated Again, C pairs with G
...GGCACGTACCTAG...
...CCGTGCATGGATC...

Over time, base changes in the DNA sequence can arise from deamination mutations. When adenine is deaminated, it becomes hypoxanthine, which can pair with cytosine. During replication, the cytosine will pair with guanine, causing an A --> G base change.

Deamination

In cpDNA, there are several A → G deamination gradients. DNA becomes susceptible to deamination events when it is single stranded. When replication forks form, the strand not being copied is single stranded, and thus at risk for A → G deamination. Therefore, gradients in deamination indicate that replication forks were most likely present and the direction that they initially opened (the highest gradient is most likely nearest the start site because it was single stranded for the longest amount of time). This mechanism is still the leading theory today; however, a second theory suggests that most cpDNA is actually linear and replicates through homologous recombination. It further contends that only a minority of the genetic material is kept in circular chromosomes while the rest is in branched, linear, or other complex structures.

Alternative Model of Replication

One of competing model for cpDNA replication asserts that most cpDNA is linear and participates in homologous recombination and replication structures similar to bacteriophage T4. It has been established that some plants have linear cpDNA, such as maize, and that more species still contain complex structures that scientists do not yet understand. When the original experiments on cpDNA were performed, scientists did notice linear structures; however, they attributed these linear forms to broken circles. If the branched and complex structures seen in cpDNA experiments are real and not artifacts of concatenated circular DNA or broken circles, then a D-loop mechanism of replication is insufficient to explain how those structures would replicate. At the same time, homologous recombination does not expand the multiple A --> G gradients seen in plastomes. Because of the failure to explain the deamination gradient as well as the numerous plant species that have been shown to have circular cpDNA, the predominant theory continues to hold that most cpDNA is circular and most likely replicates via a D loop mechanism.

Gene Content and Protein Synthesis

The chloroplast genome most commonly includes around 100 genes that code for a variety of things, mostly to do with the protein pipeline and photosynthesis. As in prokaryotes, genes in chloroplast DNA are organized into operons. Interestingly though, unlike prokaryotic DNA molecules, chloroplast DNA molecules contain introns (plant mitochondrial DNAs do too, but not human mtDNAs).

Among land plants, the contents of the chloroplast genome are fairly similar.

Chloroplast Genome Reduction and Gene Transfer

Over time, many parts of the chloroplast genome were transferred to the nuclear genome of the host, a process called *endosymbiotic gene transfer*. As a result, the chloroplast genome is heavily reduced compared to that of free-living cyanobacteria. Chloroplasts may contain 60–100 genes whereas cyanobacteria often have more than 1500 genes in their genome. Recently, a plastid without a genome was found, demonstrating chloroplasts can lose their genome during endosymbiotic gene transfer process.

Endosymbiotic gene transfer is how we know about the lost chloroplasts in many chromalveolate lineages. Even if a chloroplast is eventually lost, the genes it donated to the former host's nucleus persist, providing evidence for the lost chloroplast's existence. For example, while diatoms (a heterokontophyte) now have a red algal derived chloroplast, the presence of many green algal genes in the diatom nucleus provide evidence that the diatom ancestor (probably the ancestor of all chromalveolates too) had a green algal derived chloroplast at some point, which was subsequently replaced by the red chloroplast.

In land plants, some 11–14% of the DNA in their nuclei can be traced back to the chloroplast, up to 18% in *Arabidopsis*, corresponding to about 4,500 protein-coding genes. There have been a few recent transfers of genes from the chloroplast DNA to the nuclear genome in land plants.

Of the approximately 3000 proteins found in chloroplasts, some 95% of them are encoded by nuclear genes. Many of the chloroplast's protein complexes consist of subunits from both the chloroplast genome and the host's nuclear genome. As a result, protein synthesis must be coordinated between the chloroplast and the nucleus. The chloroplast is mostly under nuclear control, though chloroplasts can also give out signals regulating gene expression in the nucleus, called *retrograde signaling*.

Protein Synthesis

Protein synthesis within chloroplasts relies on two RNA polymerases. One is coded by the chloroplast DNA, the other is of nuclear origin. The two RNA polymerases may recognize and bind to different kinds of promoters within the chloroplast genome. The ribosomes in chloroplasts are similar to bacterial ribosomes.

Protein Targeting and Import

Because so many chloroplast genes have been moved to the nucleus, many proteins that would originally have been translated in the chloroplast are now synthesized in the cytoplasm of the plant cell. These proteins must be directed back to the chloroplast, and imported through at least two chloroplast membranes.

Curiously, around half of the protein products of transferred genes aren't even targeted back to the chloroplast. Many became exaptations, taking on new functions like participating in cell division, protein routing, and even disease resistance. A few chloroplast genes found new homes in the mitochondrial genome—most became nonfunctional pseudogenes, though a few tRNA genes still work in the mitochondrion. Some transferred chloroplast DNA protein products get directed to the secretory pathway (though it should be noted that many secondary plastids are bounded by an outermost membrane derived from the host's cell membrane, and therefore topologically outside of the cell, because to reach the chloroplast from the cytosol, you have to cross the cell membrane, just like if you were headed for the extracellular space. In those cases, chloroplast-targeted proteins do initially travel along the secretory pathway).

Because the cell acquiring a chloroplast already had mitochondria (and peroxisomes, and a cell membrane for secretion), the new chloroplast host had to develop a unique protein targeting system to avoid having chloroplast proteins being sent to the wrong organelle, and the wrong proteins being sent to the chloroplast.

The two ends of a polypeptide are called the N-terminus, or *amino end*, and the C-terminus, or *carboxyl end*. This polypeptide has four amino acids linked together. At the left is the N-terminus, with its amino (H_2N) group in green. The blue C-terminus, with its carboxyl group (CO_2H) is at the right.

In most, but not all cases, nuclear-encoded chloroplast proteins are translated with a *cleavable transit peptide* that's added to the N-terminus of the protein precursor. Sometimes the transit sequence is found on the C-terminus of the protein, or within the functional part of the protein.

Transport Proteins and Membrane Translocons

After a chloroplast polypeptide is synthesized on a ribosome in the cytosol, an enzyme specific to chloroplast proteins phosphorylates, or adds a phosphate group to many (but not all) of them in their transit sequences. Phosphorylation helps many proteins bind the polypeptide, keeping it from folding prematurely. This is important because it prevents chloroplast proteins from assuming their active form and carrying out their chloroplast functions in the wrong place—the cytosol. At the same time, they have to keep just enough shape so that they can be recognized by the chloroplast. These proteins also help the polypeptide get imported into the chloroplast.

From here, chloroplast proteins bound for the stroma must pass through two protein complexes—the TOC complex, or *translocon on the outer chloroplast membrane*, and the TIC translocon, or *translocon on the inner chloroplast membrane translocon*. Chloroplast polypeptide chains probably often travel through the two complexes at the same time, but the TIC complex can also retrieve preproteins lost in the intermembrane space.

Structure

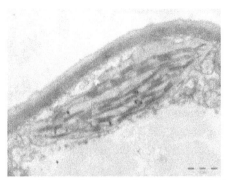

Transmission electron microscope image of a chloroplast. Grana of thylakoids and their connecting lamellae are clearly visible.

In land plants, chloroplasts are generally lens-shaped, 5–8 μm in diameter and 1–3 μm thick. Greater diversity in chloroplast shapes exists among the algae, which often contain a single chloroplast that can be shaped like a net (e.g., *Oedogonium*), a cup (e.g., *Chlamydomonas*), a ribbon-like spiral around the edges of the cell (e.g., *Spirogyra*), or slightly twisted bands at the cell edges (e.g., *Sirogonium*). Some algae have two chloroplasts in each cell; they are star-shaped in *Zygnema*, or may follow the shape of half the cell in order Desmidiales. In some algae, the chloroplast takes up most of the cell, with pockets for the nucleus and other organelles (for example some species of *Chlorella* have a cup-shaped chloroplast that occupies much of the cell).

All chloroplasts have at least three membrane systems—the outer chloroplast membrane, the inner chloroplast membrane, and the thylakoid system. Chloroplasts that are the product of secondary endosymbiosis may have additional membranes surrounding these three. Inside the outer and inner chloroplast membranes is the chloroplast stroma, a semi-gel-like fluid that makes up much of a chloroplast's volume, and in which the thylakoid system floats.

There are some common misconceptions about the outer and inner chloroplast membranes. The fact that chloroplasts are surrounded by a double membrane is often cited as evidence that they are the descendants of endosymbiotic cyanobacteria. This is often interpreted as meaning the outer chloroplast membrane is the product of the host's cell membrane infolding to form a vesicle to surround the ancestral cyanobacterium—which is not true—both chloroplast membranes are homologous to the cyanobacterium's original double membranes.

The chloroplast double membrane is also often compared to the mitochondrial double membrane. This is not a valid comparison—the inner mitochondria membrane is used to run proton pumps and carry out oxidative phosphorylation across to generate ATP energy. The only chloroplast structure that can considered analogous to it is the internal thylakoid system. Even so, in terms of "in-out", the direction of chloroplast H^+ ion flow is in the opposite direction compared to oxidative phosphorylation in mitochondria. In addition, in terms of function, the inner chloroplast membrane, which regulates metabolite passage and synthesizes some materials, has no counterpart in the mitochondrion.

Outer Chloroplast Membrane

The outer chloroplast membrane is a semi-porous membrane that small molecules and ions can easily diffuse across. However, it is not permeable to larger proteins, so chloroplast polypeptides being synthesized in the cell cytoplasm must be transported across the outer chloroplast membrane by the TOC complex, or *translocon on the outer chloroplast* membrane.

The chloroplast membranes sometimes protrude out into the cytoplasm, forming a stromule, or **strom**a-containing tub**ule**. Stromules are very rare in chloroplasts, and are much more common in other plastids like chromoplasts and amyloplasts in petals and roots, respectively. They may exist to increase the chloroplast's surface area for cross-membrane transport, because they are often branched and tangled with the endoplasmic reticulum. When they were first observed in 1962, some plant biologists dismissed the structures as artifactual, claiming that stromules were just oddly shaped chloroplasts with constricted regions or dividing chloroplasts. However, there is a growing body of evidence that stromules are functional, integral features of plant cell plastids, not merely artifacts.

Intermembrane Space and Peptidoglycan Wall

Instead of an intermembrane space, glaucophyte algae have a peptidoglycan wall between their inner and outer chloroplast membranes.

Usually, a thin intermembrane space about 10–20 nanometers thick exists between the outer and inner chloroplast membranes.

Glaucophyte algal chloroplasts have a peptidoglycan layer between the chloroplast membranes. It corresponds to the peptidoglycan cell wall of their cyanobacterial ancestors, which is located between their two cell membranes. These chloroplasts are called muroplasts (from Latin *"mura"*, meaning "wall"). Other chloroplasts have lost the cyanobacterial wall, leaving an intermembrane space between the two chloroplast envelope membranes.

Inner Chloroplast Membrane

The inner chloroplast membrane borders the stroma and regulates passage of materials in and out of the chloroplast. After passing through the TOC complex in the outer chloroplast membrane, polypeptides must pass through the TIC complex *(translocon on the inner chloroplast membrane)* which is located in the inner chloroplast membrane.

In addition to regulating the passage of materials, the inner chloroplast membrane is where fatty acids, lipids, and carotenoids are synthesized.

Peripheral Reticulum

Some chloroplasts contain a structure called the chloroplast peripheral reticulum. It is often found in the chloroplasts of C_4 plants, though it has also been found in some C_3 angiosperms, and even some gymnosperms. The chloroplast peripheral reticulum consists of a maze of membranous tubes and vesicles continuous with the inner chloroplast membrane that extends into the internal stromal fluid of the chloroplast. Its purpose is thought to be to increase the chloroplast's surface area for cross-membrane transport between its stroma and the cell cytoplasm. The small vesicles sometimes observed may serve as transport vesicles to shuttle stuff between the thylakoids and intermembrane space.

Stroma

The protein-rich, alkaline, aqueous fluid within the inner chloroplast membrane and outside of the thylakoid space is called the stroma, which corresponds to the cytosol of the original cyanobacterium. Nucleoids of chloroplast DNA, chloroplast ribosomes, the thylakoid system with plastoglobuli, starch granules, and many proteins can be found floating around in it. The Calvin cycle, which fixes CO_2 into sugar takes place in the stroma.

Chloroplast Ribosomes

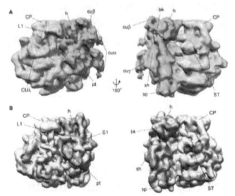

Chloroplast ribosomes Comparison of a chloroplast ribosome (green) and a bacterial ribosome (yellow). Important features common to both ribosomes and chloroplast-unique features are labeled.

Chloroplasts have their own ribosomes, which they use to synthesize a small fraction of their proteins. Chloroplast ribosomes are about two-thirds the size of cytoplasmic ribosomes (around 17 nm vs 25 nm). They take mRNAs transcribed from the chloroplast DNA and translate them into protein. While similar to bacterial ribosomes, chloroplast translation is more complex than in bacteria, so chloroplast ribosomes include some chloroplast-unique features. Small subunit ribosomal RNAs in several Chlorophyta and euglenid chloroplasts lack motifs for shine-dalgarno sequence recognition, which is considered essential for translation initiation in most chloroplasts and prokaryotes. Such loss is also rarely observed in other plastids and prokaryotes.

Plastoglobuli

Plastoglobuli (singular *plastoglobulus*, sometimes spelled *plastoglobule(s)*), are spherical bubbles of lipids and proteins about 45–60 nanometers across.They are surrounded by a lipid monolayer.Plastoglobuli are found in all chloroplasts, but become more common when the chloroplast is under oxidative stress, or when it ages and transitions into a gerontoplast. Plastoglobuli also exhibit a greater size variation under these conditions. They are also common in etioplasts, but decrease in number as the etioplasts mature into chloroplasts.

Plastoglubuli contain both structural proteins and enzymes involved in lipid synthesis and metabolism. They contain many types of lipids including plastoquinone, vitamin E, carotenoids and chlorophylls.

Plastoglobuli were once thought to be free-floating in the stroma, but it is now thought that they are permanently attached either to a thylakoid or to another plastoglobulus attached to a thylakoid, a configuration that allows a plastoglobulus to exchange its contents with the thylakoid network. In normal green chloroplasts, the vast majority of plastoglobuli occur singularly, attached directly to their parent thylakoid. In old or stressed chloroplasts, plastoglobuli tend to occur in linked groups or chains, still always anchored to a thylakoid.

Plastoglobuli form when a bubble appears between the layers of the lipid bilayer of the thylakoid membrane, or bud from existing plastoglubuli—though they never detach and float off into the stroma. Practically all plastoglobuli form on or near the highly curved edges of the thylakoid disks or sheets. They are also more common on stromal thylakoids than on granal ones.

Starch Granules

Starch granules are very common in chloroplasts, typically taking up 15% of the organelle's volume, though in some other plastids like amyloplasts, they can be big enough to distort the shape of the organelle. Starch granules are simply accumulations of starch in the stroma, and are not bounded by a membrane.

Starch granules appear and grow throughout the day, as the chloroplast synthesizes sugars, and are consumed at night to fuel respiration and continue sugar export into the phloem,though in mature chloroplasts, it is rare for a starch granule to be completely consumed or for a new granule to accumulate.

Starch granules vary in composition and location across different chloroplast lineages. In red algae, starch granules are found in the cytoplasm rather than in the chloroplast. In C_4 plants, mesophyll chloroplasts, which do not synthesize sugars, lack starch granules.

Rubisco

Rubisco, shown here in a space-filling model, is the main enzyme responsible for carbon fixation in chloroplasts.

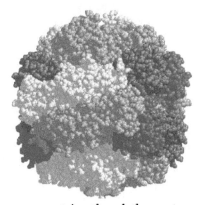

The chloroplast stroma contains many proteins, though the most common and important is Rubisco, which is probably also the most abundant protein on the planet. Rubisco is the enzyme that fixes CO_2 into sugar molecules. In C_3 plants, rubisco is abundant in all chloroplasts, though in C_4 plants, it is confined to the bundle sheath chloroplasts, where the Calvin cycle is carried out in C_4 plants.

Pyrenoids

The chloroplasts of some hornworts and algae contain structures called pyrenoids. They are not found in higher plants. Pyrenoids are roughly spherical and highly refractive bodies which are a site of starch accumulation in plants that contain them. They consist of a matrix opaque to electrons, surrounded by two hemispherical starch plates. The starch is accumulated as the pyrenoids mature. In algae with carbon concentrating mechanisms, the enzyme rubisco is found in the pyrenoids. Starch can also accumulate around the pyrenoids when CO_2 is scarce. Pyrenoids can divide to form new pyrenoids, or be produced "de novo".

Thylakoid System

Suspended within the chloroplast stroma is the thylakoid system, a highly dynamic collection of membranous sacks called thylakoids where chlorophyll is found and the light reactions of photosynthesis happen. In most vascular plant chloroplasts, the thylakoids are arranged in stacks called grana, though in certain C_4 plant chloroplasts and some algal chloroplasts, the thylakoids are free floating.

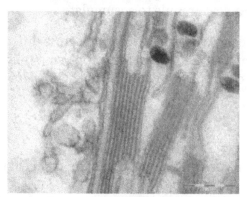

Transmission electron microscope image of some thylakoids arranged in grana stacks and lamellæ. Plastoglobuli (dark blobs) are also present.

Granal Structure

Using a light microscope, it is just barely possible to see tiny green granules—which were named grana. With electron microscopy, it became possible to see the thylakoid system in more detail, revealing it to consist of stacks of flat thylakoids which made up the grana, and long interconnecting stromal thylakoids which linked different grana. In the transmission electron microscope, thylakoid membranes appear as alternating light-and-dark bands, 8.5 nanometers thick.

For a long time, the three-dimensional structure of the thylakoid system has been unknown or disputed. One model has the granum as a stack of thylakoids linked by helical stromal thylakoids; the other has the granum as a single folded thylakoid connected in a "hub and spoke" way to other grana by stromal thylakoids. While the thylakoid system is still commonly depicted according to the folded thylakoid model, it was determined in 2011 that the stacked and helical thylakoids model is correct.

Granum structure The prevailing model for granal structure is a stack of granal thylakoids linked by helical stromal thylakoids that wrap around the grana stacks and form large sheets that connect different grana.

In the helical thylakoid model, grana consist of a stack of flattened circular granal thylakoids that resemble pancakes. Each granum can contain anywhere from two to a hundred thylakoids, though grana with 10–20 thylakoids are most common.Wrapped around the grana are helicoid stromal thylakoids, also known as frets or lamellar thylakoids. The helices ascend at an angle of 20–25°, connecting to each granal thylakoid at a bridge-like slit junction. The helicoids may extend as large sheets that link multiple grana, or narrow to tube-like bridges between grana. While different parts of the thylakoid system contain different membrane proteins, the thylakoid membranes are continuous and the thylakoid space they enclose form a single continuous labyrinth.

Thylakoids

Thylakoids (sometimes spelled *thylakoïds*), are small interconnected sacks which contain the membranes that the light reactions of photosynthesis take place on. The word *thylakoid* comes from the Greek word *thylakos* which means "sack".

Embedded in the thylakoid membranes are important protein complexes which carry out the light reactions of photosynthesis. Photosystem II and photosystem I contain light-harvesting complexes with chlorophyll and carotenoids that absorb light energy and use it to energize electrons. Molecules in the thylakoid membrane use the energized electrons to pump hydrogen ions into the thylakoid space, decreasing the pH and turning it acidic. ATP synthase is a large protein complex that harnesses the concentration gradient of the hydrogen ions in the thylakoid space to generate ATP energy as the hydrogen ions flow back out into the stroma—much like a dam turbine.

There are two types of thylakoids—granal thylakoids, which are arranged in grana, and stromal thylakoids, which are in contact with the stroma. Granal thylakoids are pancake-shaped circular disks about 300–600 nanometers in diameter. Stromal thylakoids are helicoid sheets that spiral around grana. The flat tops and bottoms of granal thylakoids contain only the relatively flat photosystem II protein complex. This allows them to stack tightly, forming grana with many layers of tightly appressed membrane, called granal membrane, increasing stability and surface area for light capture.

In contrast, photosystem I and ATP synthase are large protein complexes which jut out into the stroma. They can't fit in the appressed granal membranes, and so are found in the stromal thylakoid membrane—the edges of the granal thylakoid disks and the stromal thylakoids. These large protein complexes may act as spacers between the sheets of stromal thylakoids.

The number of thylakoids and the total thylakoid area of a chloroplast is influenced by light exposure. Shaded chloroplasts contain larger and more grana with more thylakoid membrane area than chloroplasts exposed to bright light, which have smaller and fewer grana and less thylakoid area. Thylakoid extent can change within minutes of light exposure or removal.

Pigments and Chloroplast Colors

Inside the photosystems embedded in chloroplast thylakoid membranes are various photosynthetic pigments, which absorb and transfer light energy. The types of pigments found are different in various groups of chloroplasts, and are responsible for a wide variety of chloroplast colorations.

Paper chroma-tography of some spinach leaf extract shows the various pigments present in their
chloroplasts.
Xanthophylls
Chlorophyll a
Chlorophyll b

Chlorophylls

Chlorophyll a is found in all chloroplasts, as well as their cyanobacterial ancestors.
Chlorophyll a is a blue-green pigment partially responsible for giving most cyanobacteria and chloroplasts their color. Other forms of chlorophyll exist, such as the accessory
pigments chlorophyll b, chlorophyll c, chlorophyll d, and chlorophyll f.

Chlorophyll b is an olive green pigment found only in the chloroplasts of plants, green
algae, any secondary chloroplasts obtained through the secondary endosymbiosis of a
green alga, and a few cyanobacteria. It is the chlorophylls a and b together that make
most plant and green algal chloroplasts green.

Chlorophyll c is mainly found in secondary endosymbiotic chloroplasts that originated
from a red alga, although it is not found in chloroplasts of red algae themselves. Chlorophyll c is also found in some green algae and cyanobacteria.

Chlorophylls d and f are pigments found only in some cyanobacteria.

Carotenoids

Delesseria sanguinea, a red alga, has chloroplasts that contain red pigments like phycoerytherin that
mask their blue-green chlorophyll a.

In addition to chlorophylls, another group of yellow–orange pigments called carotenoids are also found in the photosystems. There are about thirty photosynthetic carotenoids. They help transfer and dissipate excess energy, and their bright colors sometimes override the chlorophyll green, like during the fall, when the leaves of some land plants change color. β-carotene is a bright red-orange carotenoid found in nearly all chloroplasts, like chlorophyll *a*. Xanthophylls, especially the orange-red zeaxanthin, are also common. Many other forms of carotenoids exist that are only found in certain groups of chloroplasts.

Phycobilins

Phycobilins are a third group of pigments found in cyanobacteria, and glaucophyte, red algal, and cryptophyte chloroplasts. Phycobilins come in all colors, though phycoerytherin is one of the pigments that makes many red algae red. Phycobilins often organize into relatively large protein complexes about 40 nanometers across called phycobilisomes. Like photosystem I and ATP synthase, phycobilisomes jut into the stroma, preventing thylakoid stacking in red algal chloroplasts. Cryptophyte chloroplasts and some cyanobacteria don't have their phycobilin pigments organized into phycobilisomes, and keep them in their thylakoid space instead.

Photosynthetic pigments Table of the presence of various pigments across chloroplast groups. Colored cells represent pigment presence.

Specialized Chloroplasts in C$_4$ Plants

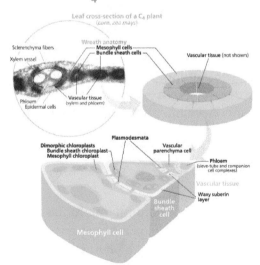

Many C$_4$ plants have their mesophyll cells and bundle sheath cells arranged radially around their leaf veins. The two types of cells contain different types of chloroplasts specialized for a particular part of photosynthesis.

To fix carbon dioxide into sugar molecules in the process of photosynthesis, chloroplasts use an enzyme called rubisco. Rubisco has a problem—it has trouble distinguishing between carbon dioxide and oxygen, so at high oxygen concentrations, rubisco starts accidentally adding oxygen to sugar precursors. This has the end result of ATP energy being wasted and CO_2 being released, all with no sugar being produced. This is a big problem, since O_2 is produced by the initial light reactions of photosynthesis, causing issues down the line in the Calvin cycle which uses rubisco.

C4 plants evolved a way to solve this—by spatially separating the light reactions and the Calvin cycle. The light reactions, which store light energy in ATP and NADPH, are done in the mesophyll cells of a C_4 leaf. The Calvin cycle, which uses the stored energy to make sugar using rubisco, is done in the bundle sheath cells, a layer of cells surrounding a vein in a leaf.

As a result, chloroplasts in C_4 mesophyll cells and bundle sheath cells are specialized for each stage of photosynthesis. In mesophyll cells, chloroplasts are specialized for the light reactions, so they lack rubisco, and have normal grana and thylakoids, which they use to make ATP and NADPH, as well as oxygen. They store CO_2 in a four-carbon compound, which is why the process is called C_4 photosynthesis. The four-carbon compound is then transported to the bundle sheath chloroplasts, where it drops off CO_2 and returns to the mesophyll. Bundle sheath chloroplasts do not carry out the light reactions, preventing oxygen from building up in them and disrupting rubisco activity. Because of this, they lack thylakoids organized into grana stacks—though bundle sheath chloroplasts still have free-floating thylakoids in the stroma where they still carry out cyclic electron flow, a light-driven method of synthesizing ATP to power the Calvin cycle without generating oxygen. They lack photosystem II, and only have photosystem I—the only protein complex needed for cyclic electron flow. Because the job of bundle sheath chloroplasts is to carry out the Calvin cycle and make sugar, they often contain large starch grains.

Both types of chloroplast contain large amounts of chloroplast peripheral reticulum, which they use to get more surface area to transport stuff in and out of them. Mesophyll chloroplasts have a little more peripheral reticulum than bundle sheath chloroplasts.

Location

Distribution in a Plant

Not all cells in a multicellular plant contain chloroplasts. All green parts of a plant contain chloroplasts—the chloroplasts, or more specifically, the chlorophyll in them are what make the photosynthetic parts of a plant green. The plant cells which contain chloroplasts are usually parenchyma cells, though chloroplasts can also be found in collenchyma tissue. A plant cell which contains chloroplasts is known as a chlorenchyma cell. A typical chlorenchyma cell of a land plant contains about 10 to 100 chloroplasts.

In some plants such as cacti, chloroplasts are found in the stems, though in most plants, chloroplasts are concentrated in the leaves. One square millimeter of leaf tissue can contain half a million chloroplasts. Within a leaf, chloroplasts are mainly found in the mesophyll layers of a leaf, and the guard cells of stomata. Palisade mesophyll cells can contain 30–70 chloroplasts per cell, while stomatal guard cells contain only around 8–15 per cell, as well as much less chlorophyll. Chloroplasts can also be found in the bundle sheath cells of a leaf, especially in C_4 plants, which carry out the Calvin cycle in their bundle sheath cells. They are often absent from the epidermis of a leaf.

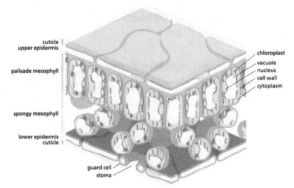

A cross section of a leaf, showing chloroplasts in its mesophyll cells. Stomal guard cells also have chloroplasts, though much fewer than mesophyll cells.

Cellular Location

Chloroplast Movement

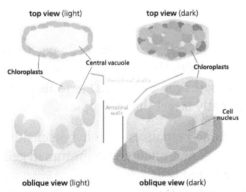

When chloroplasts are exposed to direct sunlight, they stack along the anticlinal cell walls to minimize exposure. In the dark they spread out in sheets along the periclinal walls to maximize light absorption.

The chloroplasts of plant and algal cells can orient themselves to best suit the available light. In low-light conditions, they will spread out in a sheet—maximizing the surface area to absorb light. Under intense light, they will seek shelter by aligning in vertical columns along the plant cell's cell wall or turning sideways so that light strikes them edge-on. This reduces exposure and protects them from photooxidative damage. This ability to distribute chloroplasts so that they can take shelter behind each other or spread out may be the reason why land plants evolved to have many small chloroplasts instead of a few big ones. Chloroplast movement is considered one of the most closely regulated stimulus-response systems that can be found in plants. Mitochondria have also been observed to follow chloroplasts as they move.

In higher plants, chloroplast movement is run by phototropins, blue light photoreceptors also responsible for plant phototropism. In some algae, mosses, ferns, and flowering plants, chloroplast movement is influenced by red light in addition to blue light,though very long red wavelengths inhibit movement rather than speeding it up. Blue light generally causes chloroplasts to seek shelter, while red light draws them out to maximize light absorption.

Studies of *Vallisneria gigantea*, an aquatic flowering plant, have shown that chloroplasts can get moving within five minutes of light exposure, though they don't initially show any net directionality. They may move along microfilament tracks, and the fact that the microfilament mesh changes shape to form a honeycomb structure surrounding the chloroplasts after they have moved suggests that microfilaments may help to anchor chloroplasts in place.

Function and Chemistry

Guard Cell Chloroplasts

Unlike most epidermal cells, the guard cells of plant stomata contain relatively well-developed chloroplasts. However, exactly what they do is controversial.

Plant Innate Immunity

Plants lack specialized immune cells—all plant cells participate in the plant immune response. Chloroplasts, along with the nucleus, cell membrane, and endoplasmic reticulum, are key players in pathogen defense. Due to its role in a plant cell's immune response, pathogens frequently target the chloroplast.

Plants have two main immune responses—the hypersensitive response, in which infected cells seal themselves off and undergo programmed cell death, and systemic acquired resistance, where infected cells release signals warning the rest of the plant of a pathogen's presence. Chloroplasts stimulate both responses by purposely damaging their photosynthetic system, producing reactive oxygen species. High levels of reactive oxygen species will cause the hypersensitive response. The reactive oxygen species also directly kill any pathogens within the cell. Lower levels of reactive oxygen species initiate systemic acquired resistance, triggering defense-molecule production in the rest of the plant.

In some plants, chloroplasts are known to move closer to the infection site and the nucleus during an infection.

Chloroplasts can serve as cellular sensors. After detecting stress in a cell, which might be due to a pathogen, chloroplasts begin producing molecules like salicylic acid, jasmonic acid, nitric oxide and reactive oxygen species which can serve as defense-signals. As cellular signals, reactive oxygen species are unstable molecules, so they probably don't leave the chloroplast, but instead pass on their signal to an unknown second messenger molecule. All these molecules initiate retrograde signaling—signals from the chloroplast that regulate gene expression in the nucleus.

In addition to defense signaling, chloroplasts, with the help of the peroxisomes, help synthesize an important defense molecule, jasmonate. Chloroplasts synthesize all the fatty acids in a plant cell—linoleic acid, a fatty acid, is a precursor to jasmonate.

Photosynthesis

One of the main functions of the chloroplast is its role in photosynthesis, the process by which light is transformed into chemical energy, to subsequently produce food in the form of sugars. Water (H_2O) and carbon dioxide (CO_2) are used in photosynthesis, and sugar and oxygen (O_2) is made, using light energy. Photosynthesis is divided into two stages—the light reactions, where water is split to produce oxygen, and the dark reactions, or Calvin cycle, which builds sugar molecules from carbon dioxide. The two phases are linked by the energy carriers adenosine triphosphate (ATP) and nicotinamide adenine dinucleotide phosphate ($NADP^+$).

Light Reactions

The light reactions of photosynthesis take place across the thylakoid membranes.

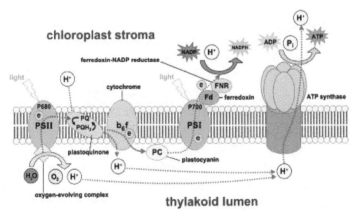

The light reactions take place on the thylakoid membranes. They take light energy and store it in NADPH, a form of $NADP^+$, and ATP to fuel the dark reactions.

Energy Carriers

ATP is the phosphorylated version of adenosine diphosphate (ADP), which stores energy in a cell and powers most cellular activities. ATP is the energized form, while ADP is the (partially) depleted form. $NADP^+$ is an electron carrier which ferries high energy electrons. In the light reactions, it gets reduced, meaning it picks up electrons, becoming NADPH.

Photophosphorylation

Like mitochondria, chloroplasts use the potential energy stored in an H^+, or hydrogen ion gradient to generate ATP energy. The two photosystems capture light energy to energize electrons taken from water, and release them down an electron transport chain. The molecules between the photosystems harness the electrons' energy to pump hydrogen ions into the thylakoid space, creating a concentration gradient, with more hydrogen ions (up to a thousand times as many) inside the thylakoid system than in the stroma. The hydrogen ions in the thylakoid space then diffuse back down their concentration gradient, flowing back out into the stroma through ATP synthase. ATP synthase uses the energy from the flowing hydrogen ions to phosphorylate adenosine diphosphate into adenosine triphosphate, or ATP. Because chloroplast ATP synthase projects out into the stroma, the ATP is synthesized there, in position to be used in the dark reactions.

$NADP^+$ Reduction

Electrons are often removed from the electron transport chains to charge $NADP^+$ with electrons, reducing it to NADPH. Like ATP synthase, ferredoxin-$NADP^+$ reductase, the enzyme that reduces $NADP^+$, releases the NADPH it makes into the stroma, right where it is needed for the dark reactions.

Because $NADP^+$ reduction removes electrons from the electron transport chains, they must be replaced—the job of photosystem II, which splits water molecules (H_2O) to obtain the electrons from its hydrogen atoms.

Cyclic Photophosphorylation

While photosystem II photolyzes water to obtain and energize new electrons, photosystem I simply reenergizes depleted electrons at the end of an electron transport chain. Normally, the reenergized electrons are taken by $NADP^+$, though sometimes they can flow back down more H^+-pumping electron transport chains to transport more hydrogen ions into the thylakoid space to generate more ATP. This is termed cyclic photophosphorylation because the electrons are recycled. Cyclic photophosphorylation is common in C_4 plants, which need more ATP than NADPH.

Dark Reactions

The Calvin cycle (Interactive diagram) The Calvin cycle incorporates carbon dioxide into sugar molecules.

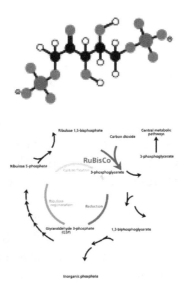

RuBisCo

The Calvin cycle, also known as the dark reactions, is a series of biochemical reactions that fixes CO_2 into G3P sugar molecules and uses the energy and electrons from the ATP and NADPH made in the light reactions. The Calvin cycle takes place in the stroma of the chloroplast.

While named *"the dark reactions"*, in most plants, they take place in the light, since the dark reactions are dependent on the products of the light reactions.

Carbon Fixation and G3P Synthesis

The Calvin cycle starts by using the enzyme Rubisco to fix CO_2 into five-carbon Ribulose bisphosphate (RuBP) molecules. The result is unstable six-carbon molecules that immediately break down into three-carbon molecules called 3-phosphoglyceric acid, or 3-PGA. The ATP and NADPH made in the light reactions is used to convert the 3-PGA into glyceraldehyde-3-phosphate, or G3P sugar molecules. Most of the G3P molecules are recycled back into RuBP using energy from more ATP, but one out of every six produced leaves the cycle—the end product of the dark reactions.

Sugars and Starches

Glyceraldehyde-3-phosphate can double up to form larger sugar molecules like glucose and fructose. These molecules are processed, and from them, the still larger sucrose, a disaccharide commonly known as table sugar, is made, though this process takes place outside of the chloroplast, in the cytoplasm.

Sucrose is made up of a glucose monomer (left), and a fructose monomer (right).

Alternatively, glucose monomers in the chloroplast can be linked together to make starch, which accumulates into the starch grains found in the chloroplast. Under conditions such as high atmospheric CO_2 concentrations, these starch grains may grow very large, distorting the grana and thylakoids. The starch granules displace the thylakoids, but leave them intact. Waterlogged roots can also cause starch buildup in the chloroplasts, possibly due to less sucrose being exported out of the chloroplast (or more accurately, the plant cell). This depletes a plant's free phosphate supply, which indirectly stimulates chloroplast starch synthesis. While linked to low photosynthesis rates, the starch grains themselves may not necessarily interfere significantly with the efficiency of photosynthesis, and might simply be a side effect of another photosynthesis-depressing factor.

Photorespiration

Photorespiration can occur when the oxygen concentration is too high. Rubisco cannot distinguish between oxygen and carbon dioxide very well, so it can accidentally add O_2 instead of CO_2 to RuBP. This process reduces the efficiency of photosynthesis—it consumes ATP and oxygen, releases CO_2, and produces no sugar. It can waste up to half the carbon fixed by the Calvin cycle. Several mechanisms have evolved in different lineages that raise the carbon dioxide concentration relative to oxygen within the chloroplast, increasing the efficiency of photosynthesis. These mechanisms are called carbon dioxide concentrating mechanisms, or CCMs. These include Crassulacean acid metabolism, C_4 carbon fixation, and pyrenoids. Chloroplasts in C_4 plants are notable as they exhibit a distinct chloroplast dimorphism.

pH

Because of the H^+ gradient across the thylakoid membrane, the interior of the thylakoid is acidic, with a pH around 4, while the stroma is slightly basic, with a pH of around 8. The optimal stroma pH for the Calvin cycle is 8.1, with the reaction nearly stopping when the pH falls below 7.3.

$CO2$ in water can form carbonic acid, which can disturb the pH of isolated chloroplasts, interfering with photosynthesis, even though CO_2 is used in photosynthesis. However, chloroplasts in living plant cells are not affected by this as much.

Chloroplasts can pump K^+ and H^+ ions in and out of themselves using a poorly understood light-driven transport system.

In the presence of light, the pH of the thylakoid lumen can drop up to 1.5 pH units, while the pH of the stroma can rise by nearly one pH unit.

Amino Acid Synthesis

Chloroplasts alone make almost all of a plant cell's amino acids in their stroma except the sulfur-containing ones like cysteine and methionine.Cysteine is made in the chloroplast (the proplastid too) but it is also synthesized in the cytosol and mitochondria, probably because it has trouble crossing membranes to get to where it is needed. The chloroplast is known to make the precursors to methionine but it is unclear whether the organelle carries out the last leg of the pathway or if it happens in the cytosol.

Other Nitrogen Compounds

Chloroplasts make all of a cell's purines and pyrimidines—the nitrogenous bases found in DNA and RNA. They also convert nitrite (NO_2^-) into ammonia (NH_3) which supplies the plant with nitrogen to make its amino acids and nucleotides.

Other Chemical Products

Chloroplasts are the site of complex lipid metabolism.

Differentiation, Replication, and Inheritance

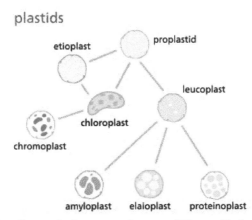

Plastid types (Interactive diagram) Plants contain many different kinds of plastids in their cells.

Chloroplasts are a special type of a plant cell organelle called a plastid, though the two terms are sometimes used interchangeably. There are many other types of plastids, which carry out various functions. All chloroplasts in a plant are descended from undifferentiated proplastids found in the zygote, or fertilized egg. Proplastids are commonly found in an adult plant's apical meristems. Chloroplasts do not normally develop from proplastids in root tip meristems—instead, the formation of starch-storing amyloplasts is more common.

In shoots, proplastids from shoot apical meristems can gradually develop into chloro-plasts in photosynthetic leaf tissues as the leaf matures, if exposed to the required light. This process involves invaginations of the inner plastid membrane, forming sheets of membrane that project into the internal stroma. These membrane sheets then fold to form thylakoids and grana.

If angiosperm shoots are not exposed to the required light for chloroplast formation, proplastids may develop into an etioplast stage before becoming chloroplasts. An etio-plast is a plastid that lacks chlorophyll, and has inner membrane invaginations that form a lattice of tubes in their stroma, called a prolamellar body. While etioplasts lack chlorophyll, they have a yellow chlorophyll precursor stocked. Within a few minutes of light exposure, the prolamellar body begins to reorganize into stacks of thylakoids, and chlorophyll starts to be produced. This process, where the etioplast becomes a chloro-plast, takes several hours. Gymnosperms do not require light to form chloroplasts.

Light, however, does not guarantee that a proplastid will develop into a chloroplast. Whether a proplastid develops into a chloroplast some other kind of plastid is mostly controlled by the nucleus and is largely influenced by the kind of cell it resides in.

possible plastid interconversions

Many plastid interconversions are possible.

Plastid Interconversion

Plastid differentiation is not permanent, in fact many interconversions are possible. Chloroplasts may be converted to chromoplasts, which are pigment-filled plastids re-sponsible for the bright colors seen in flowers and ripe fruit. Starch storing amyloplasts can also be converted to chromoplasts, and it is possible for proplastids to develop straight into chromoplasts. Chromoplasts and amyloplasts can also become chloro-plasts, like what happens when a carrot or a potato is illuminated. If a plant is injured, or something else causes a plant cell to revert to a meristematic state, chloroplasts and other plastids can turn back into proplastids. Chloroplast, amyloplast, chromoplast, proplast, etc., are not absolute states—intermediate forms are common.

Chloroplast Division

Most chloroplasts in a photosynthetic cell do not develop directly from proplastids or etio-plasts. In fact, a typical shoot meristematic plant cell contains only 7–20 proplastids. These proplastids differentiate into chloroplasts, which divide to create the 30–70 chloroplasts found in a mature photosynthetic plant cell. If the cell divides, chloroplast division provides the additional chloroplasts to partition between the two daughter cells.

In single-celled algae, chloroplast division is the only way new chloroplasts are formed. There is no proplastid differentiation—when an algal cell divides, its chloroplast divides along with it, and each daughter cell receives a mature chloroplast.

Almost all chloroplasts in a cell divide, rather than a small group of rapidly dividing chloroplasts. Chloroplasts have no definite S-phase—their DNA replication is not syn-chronized or limited to that of their host cells. Much of what we know about chloroplast division comes from studying organisms like *Arabidopsis* and the red alga *Cyanid-ioschyzon merolæ*.

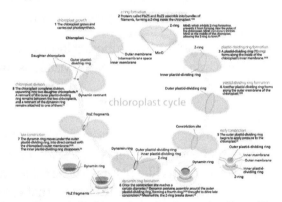

Most chloroplasts in plant cells, and all chloroplasts in algae arise from chloroplast division. *Picture references,*

The division process starts when the proteins FtsZ1 and FtsZ2 assemble into filaments, and with the help of a protein ARC6, form a structure called a Z-ring within the chloroplast's stroma. The Min system manages the placement of the Z-ring, ensuring that the chloroplast is cleaved more or less evenly. The protein MinD prevents FtsZ from linking up and forming filaments. Another protein ARC3 may also be involved, but it is not very well understood. These proteins are active at the poles of the chloroplast, preventing Z-ring formation there, but near the center of the chloroplast, MinE inhibits them, allowing the Z-ring to form.

Next, the two plastid-dividing rings, or PD rings form. The inner plastid-dividing ring is located in the inner side of the chloroplast's inner membrane, and is formed first. The outer plastid-dividing ring is found wrapped around the outer chloroplast membrane. It con-sists of filaments about 5 nanometers across, arranged in rows 6.4 nanometers apart, and shrinks to squeeze the chloroplast. This is when chloroplast constriction begins.

In a few species like *Cyanidioschyzon merolæ*, chloroplasts have a third plastid-dividing ring located in the chloroplast's intermembrane space.

Late into the constriction phase, dynamin proteins assemble around the outer plastid-dividing ring, helping provide force to squeeze the chloroplast.Meanwhile, the Z-ring and the inner plastid-dividing ring break down. During this stage, the many chloroplast DNA plasmids floating around in the stroma are partitioned and distributed to the two forming daughter chloroplasts.

Later, the dynamins migrate under the outer plastid dividing ring, into direct contact with the chloroplast's outer membrane, to cleave the chloroplast in two daughter chloroplasts.

A remnant of the outer plastid dividing ring remains floating between the two daughter chloroplasts, and a remnant of the dynamin ring remains attached to one of the daughter chloroplasts.

Of the five or six rings involved in chloroplast division, only the outer plastid-dividing ring is present for the entire constriction and division phase—while the Z-ring forms first, constriction does not begin until the outer plastid-dividing ring forms.

Chloroplast division In this light micrograph of some moss chloroplasts, many dumbbell-shaped chloroplasts can be seen dividing. Grana are also just barely visible as small granules.

Regulation

In species of algae that contain a single chloroplast, regulation of chloroplast division is extremely important to ensure that each daughter cell receives a chloroplast—chloroplasts can't be made from scratch. In organisms like plants, whose cells contain multiple chloroplasts, coordination is looser and less important. It is likely that chloroplast and cell division are somewhat synchronized, though the mechanisms for it are mostly unknown.

Light has been shown to be a requirement for chloroplast division. Chloroplasts can grow and progress through some of the constriction stages under poor quality green light, but are slow to complete division—they require exposure to bright white light to complete division. Spinach leaves grown under green light have been observed to contain many large dumbbell-shaped chloroplasts. Exposure to white light can stimulate these chloroplasts to divide and reduce the population of dumbbell-shaped chloroplasts.

Chloroplast Inheritance

Like mitochondria, chloroplasts are usually inherited from a single parent. Biparental chloroplast inheritance—where plastid genes are inherited from both parent plants—occurs in very low levels in some flowering plants.

Many mechanisms prevent biparental chloroplast DNA inheritance, including selective destruction of chloroplasts or their genes within the gamete or zygote, and chloroplasts from one parent being excluded from the embryo. Parental chloroplasts can be sorted so that only one type is present in each offspring.

Gymnosperms, such as pine trees, mostly pass on chloroplasts paternally, while flowering plants often inherit chloroplasts maternally. Flowering plants were once thought to only inherit chloroplasts maternally. However, there are now many documented cases of angiosperms inheriting chloroplasts paternally.

Angiosperms, which pass on chloroplasts maternally, have many ways to prevent paternal inheritance. Most of them produce sperm cells that do not contain any plastids. There are many other documented mechanisms that prevent paternal inheritance in these flowering plants, such as different rates of chloroplast replication within the embryo.

Among angiosperms, paternal chloroplast inheritance is observed more often in hybrids than in offspring from parents of the same species. This suggests that incompatible hybrid genes might interfere with the mechanisms that prevent paternal inheritance.

Transplastomic Plants

Recently, chloroplasts have caught attention by developers of genetically modified crops. Since, in most flowering plants, chloroplasts are not inherited from the male parent, transgenes in these plastids cannot be disseminated by pollen. This makes plastid transformation a valuable tool for the creation and cultivation of genetically modified plants that are biologically contained, thus posing significantly lower environmental risks. This biological containment strategy is therefore suitable for establishing the coexistence of conventional and organic agriculture. While the reliability of this mechanism has not yet been studied for all relevant crop species, recent results in tobacco plants are promising, showing a failed containment rate of transplastomic plants at 3 in 1,000,000.

Thylakoid

Thylakoids (dark green) inside a chloroplast

A thylakoid is a membrane-bound compartment inside chloroplasts and cyanobacteria. They are the site of the light-dependent reactions of photosynthesis. Thylakoids consist of a thylakoid membrane surrounding a thylakoid lumen. Chloroplast thylakoids frequently form stacks of disks referred to as grana (singular: granum). Grana are connected by intergranal or stroma thylakoids, which join granum stacks together as a single functional compartment.

Thylakoid Structure

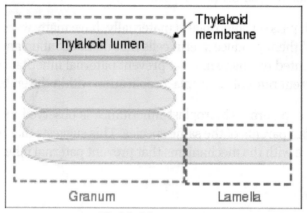

Thylakoid structures

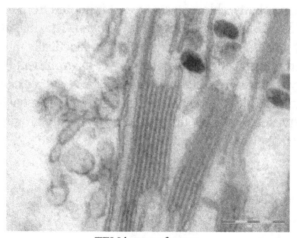

TEM image of grana

Thylakoids are membrane-bound structures embedded in the chloroplast stroma. A stack of thylakoids is called a granum and resembles a stack of coins.

Membrane

The thylakoid membrane is the site of the light-dependent reactions of photosynthesis with the photosynthetic pigments embedded directly in the membrane. It is an alternating pattern of dark and light bands measuring each 1 nanometre. The thylakoid lipid bilayer shares characteristic features with prokaryotic membranes and the inner chloroplast membrane. For example, acidic lipids can be found in thylakoid membranes, cyanobacteria and other photosynthetic bacteria and are involved in the functional integrity of the photosystems. The thylakoid membranes of higher plants are composed primarily of phospholipids and galactolipids that are asymmetrically arranged along and across the membranes. Thylakoid membranes are richer in galactolipids rather than phospholipids; also they predominantly consist of hexagonal phase II forming monogalacotosyl diglyceride lipid. Despite this unique composition, plant thylakoid membranes have been shown to assume largely lipid-bilayer dynamic organization. Lipids forming the thylakoid membranes, richest in high-fluidity linolenic acid are synthesized in a complex pathway involving exchange of lipid precursors between the endoplasmic reticulum and inner membrane of the plastid envelope and transported from the inner membrane to the thylakoids via vesicles.

Lumen

The thylakoid lumen is a continuous aqueous phase enclosed by the thylakoid membrane. It plays an important role for photophosphorylation during photosynthesis. During the light-dependent reaction, protons are pumped across the thylakoid membrane into the lumen making it acidic down to pH 4.

Granum and Stroma Lamellae

In higher plants thylakoids are organized into a granum-stroma membrane assembly. A granum (plural grana) is a stack of thylakoid discs. Chloroplasts can have from 10 to 100 grana. Grana are connected by stroma thylakoids, also called intergranal thylakoids or lamellae. Grana thylakoids and stroma thylakoids can be distinguished by their different protein composition. Grana contribute to chloroplasts' large surface area to volume ratio. Different interpretations of electron tomography imaging of thylakoid membranes has resulted in two models for grana structure. Both posit that lamellae intersect grana stacks in parallel sheets, though whether these sheets intersect in planes perpendicular to the grana stack axis, or are arranged in a right-handed helix is debated.

Thylakoid Formation

Chloroplasts develop from proplastids when seedlings emerge from the ground. Thylakoid formation requires light. In the plant embryo and in the absence of light, proplastids develop into etioplasts that contain semicrystalline membrane structures called prolamellar bodies. When exposed to light, these prolamellar bodies develop into thylakoids. This does not happen in seedlings grown in the dark, which undergo etiolation. An underexposure to light can cause the thylakoids to fail. This causes the chloroplasts to fail resulting in the death of the plant.

Thylakoid formation requires the action of *vesicle-inducing protein in plastids 1* (VIPP1). Plants cannot survive without this protein, and reduced VIPP1 levels lead to slower growth and paler plants with reduced ability to photosynthesize. VIPP1 appears to be required for basic thylakoid membrane formation, but not for the assembly of protein complexes of the thylakoid membrane. It is conserved in all organisms containing thylakoids, including cyanobacteria, green algae, such as Chlamydomonas, and higher plants, such as *Arabidopsis thaliana.*

Thylakoid Isolation and Fractionation

Thylakoids can be purified from plant cells using a combination of differential and gradient centrifugation. Disruption of isolated thylakoids, for example by mechanical shearing, releases the lumenal fraction. Peripheral and integral membrane fractions can be extracted from the remaining membrane fraction. Treatment with sodium carbonate (Na_2CO_3) detaches peripheral membrane proteins, whereas treatment with detergents and organic solvents solubilizes integral membrane proteins.

Thylakoid Proteins

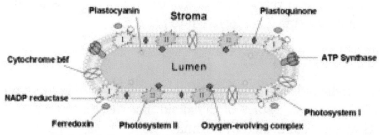

Thylakoid disc with embedded and associated proteins

Thylakoids contain many integral and peripheral membrane proteins, as well as lumenal proteins. Recent proteomics studies of thylakoid fractions have provided further details on the protein composition of the thylakoids. These data have been summarized in several plastid protein databases that are available online.

According to these studies, the thylakoid proteome consists of at least 335 different proteins. Out of these, 89 are in the lumen, 116 are integral membrane proteins, 62 are peripheral proteins on the stroma side, and 68 peripheral proteins on the lumenal side. Additional low-abundance lumenal proteins can be predicted through computational methods. Of the thylakoid proteins with known functions, 42% are involved in photosynthesis. The next largest functional groups include proteins involved in protein targeting, processing and folding with 11%, oxidative stress response (9%) and translation (8%).

Integral Membrane Proteins

Thylakoid membranes contain integral membrane proteins which play an important role in light harvesting and the light-dependent reactions of photosynthesis. There are four major protein complexes in the thylakoid membrane:

- Photosystems I and II

- Cytochrome b6f complex

- ATP synthase

Photosystem II is located mostly in the grana thylakoids, whereas photosystem I and ATP synthase are mostly located in the stroma thylakoids and the outer layers of grana. The cytochrome b6f complex is distributed evenly throughout thylakoid membranes. Due to the separate location of the two photosystems in the thylakoid membrane system, mobile electron carriers are required to shuttle electrons between them. These carriers are plastoquinone and plastocyanin. Plastoquinone shuttles electrons from photosystem II to the cytochrome b6f complex, whereas plastocyanin carries electrons from the cytochrome b6f complex to photosystem I.

Together, these proteins make use of light energy to drive electron transport chains that generate a chemiosmotic potential across the thylakoid membrane and NADPH, a product of the terminal redox reaction. The ATP synthase uses the chemiosmotic potential to make ATP during photophosphorylation.

Photosystems

These photosystems are light-driven redox centers, each consisting of an antenna complex that uses chlorophylls and accessory photosynthetic pigments such as carotenoids and phycobiliproteins to harvest light at a variety of wavelengths. Each antenna complex has between 250 and 400 pigment molecules and the energy they absorb is shuttled by resonance energy transfer to a specialized chlorophyll a at the reaction center of each photosystem. When either of the two chlorophyll a molecules at the reaction center absorbs energy, an electron is excited and transferred to an electron-acceptor molecule. Photosystem I contains a pair of chlorophyll a molecules, designated P700, at its reaction center that maximally absorbs 700 nm light. Photosystem II contains P680 chlorophyll that absorbs 680 nm light best (note that these wavelengths correspond to deep red - see the visible spectrum). The P is short for pigment and the number is the specific absorption peak in nanometers for the chlorophyll molecules in each reaction center.

Cytochrome b6f Complex

The cytochrome b6f complex is part of the thylakoid electron transport chain and couples electron transfer to the pumping of protons into the thylakoid lumen. Energetically, it is situated between the two photosystems and transfers electrons from photosystem II-plastoquinone to plastocyanin-photosystem I.

ATP Synthase

The thylakoid ATP synthase is a CF1FO-ATP synthase similar to the mitochondrial AT-

Pase. It is integrated into the thylakoid membrane with the CF1-part sticking into stroma. Thus, ATP synthesis occurs on the stromal side of the thylakoids where the ATP is needed for the light-independent reactions of photosynthesis.

Thylakoid Lumen Proteins

The electron transport protein plastocyanin is present in the lumen and shuttles electrons from the cytochrome b6f protein complex to photosystem I. While plastoquinones are lipid-soluble and therefore move within the thylakoid membrane, plastocyanin moves through the thylakoid lumen.

The lumen of the thylakoids is also the site of water oxidation by the oxygen evolving complex associated with the lumenal side of photosystem II.

Lumenal proteins can be predicted computationally based on their targeting signals. In Arabidopsis, out of the predicted lumenal proteins possessing the Tat signal, the largest groups with known functions are 19% involved in protein processing (proteolysis and folding), 18% in photosynthesis, 11% in metabolism, and 7% redox carriers and defense.

Thylakoid Protein Expression

Chloroplasts have their own genome, which encodes a number of thylakoid proteins. However, during the course of plastid evolution from their cyanobacterial endosymbiotic ancestors, extensive gene transfer from the chloroplast genome to the cell nucleus took place. This results in the four major thylakoid protein complexes being encoded in part by the chloroplast genome and in part by the nuclear genome. Plants have developed several mechanisms to co-regulate the expression of the different subunits encoded in the two different organelles to assure the proper stoichiometry and assembly of these protein complexes. For example, transcription of nuclear genes encoding parts of the photosynthetic apparatus is regulated by light. Biogenesis, stability and turnover of thylakoid protein complexes are regulated by phosphorylation via redox-sensitive kinases in the thylakoid membranes. The translation rate of chloroplast-encoded proteins is controlled by the presence or absence of assembly partners (control by epistasy of synthesis). This mechanism involves negative feedback through binding of excess protein to the 5' untranslated region of the chloroplast mRNA. Chloroplasts also need to balance the ratios of photosystem I and II for the electron transfer chain. The redox state of the electron carrier plastoquinone in the thylakoid membrane directly affects the transcription of chloroplast genes encoding proteins of the reaction centers of the photosystems, thus counteracting imbalances in the electron transfer chain.

Protein Targeting to the Thylakoids

Thylakoid proteins are targeted to their destination via signal peptides and prokaryotic-type secretory pathways inside the chloroplast. Most thylakoid proteins encoded by a plant's nuclear genome need two targeting signals for proper localization: An N-terminal chloroplast targeting peptide (shown in yellow in the figure), followed by a thylakoid targeting peptide (shown in blue). Proteins are imported through the translocon of outer and inner membrane (Toc and Tic) complexes. After entering the chloroplast, the first targeting peptide is cleaved off by a protease processing imported proteins. This unmasks the second targeting signal and the protein is exported from the stroma into the thylakoid in a second targeting step. This second step requires the action of protein translocation components of the thylakoids and is energy-dependent. Proteins are inserted into the membrane via the SRP-dependent pathway (1), the Tat-dependent pathway (2), or spontaneously via their transmembrane domains (not shown in figure). Lumenal proteins are exported across the thylakoid membrane into the lumen by either the Tat-dependent pathway (2) or the Sec-dependent pathway (3) and released by cleavage from the thylakoid targeting signal. The different pathways utilize different signals and energy sources. The Sec (secretory) pathway requires ATP as energy source and consists of SecA, which binds to the imported protein and a Sec membrane complex to shuttle the protein across. Proteins with a twin arginine motif in their thylakoid signal peptide are shuttled through the Tat (twin arginine translocation) pathway, which requires a membrane-bound Tat complex and the pH gradient as an energy source. Some other proteins are inserted into the membrane via the SRP (signal recognition particle) pathway. The chloroplast SRP can interact with its target proteins either post-translationally or co-translationally, thus transporting imported proteins as well as those that are translated inside the chloroplast. The SRP pathway requires GTP and the pH gradient as energy sources. Some transmembrane proteins may also spontaneously insert into the membrane from the stromal side without energy requirement.

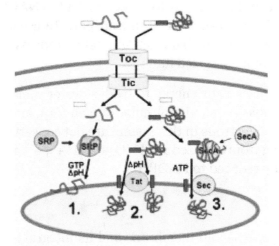

Schematic representation of thylakoid protein targeting pathways.

Thylakoid Function

The thylakoids are the site of the light-dependent reactions of photosynthesis. These include light-driven water oxidation and oxygen evolution, the pumping of protons across the thylakoid membranes coupled with the electron transport chain of the photosystems and cytochrome complex, and ATP synthesis by the ATP synthase utilizing the generated proton gradient.

Water Photolysis

The first step in photosynthesis is the light-driven reduction (splitting) of water to provide the electrons for the photosynthetic electron transport chains as well as protons for the establishment of a proton gradient. The water-splitting reaction occurs on the lumenal side of the thylakoid membrane and is driven by the light energy captured by the photosystems. It is interesting to note that this oxidation of water conveniently produces the waste product O_2 that is vital for cellular respiration. The molecular oxygen formed by the reaction is released into the atmosphere.

Electron Transport Chains

Two different variations of electron transport are used during photosynthesis:

- Noncyclic electron transport or Non-cyclic photophosphorylation produces NADPH + H^+ and ATP.

- Cyclic electron transport or Cyclic photophosphorylation produces only ATP.

The noncyclic variety involves the participation of both photosystems, while the cyclic electron flow is dependent on only photosystem I.

- Photosystem I uses light energy to reduce $NADP^+$ to NADPH + H^+, and is active in both noncyclic and cyclic electron transport. In cyclic mode, the energized electron is passed down a chain that ultimately returns it (in its base state) to the chlorophyll that energized it.

- Photosystem II uses light energy to oxidize water molecules, producing electrons (e^-), protons (H^+), and molecular oxygen (O_2), and is only active in noncyclic transport. Electrons in this system are not conserved, but are rather continually entering from oxidized $2H_2O$ (O_2 + 4 H^+ + 4 e^-) and exiting with $NADP^+$ when it is finally reduced to NADPH.

Chemiosmosis

A major function of the thylakoid membrane and its integral photosystems is the establishment of chemiosmotic potential. The carriers in the electron transport chain use some of the electron's energy to actively transport protons from the stroma to the lumen. During photosynthesis, the lumen becomes acidic, as low as pH 4, compared to pH 8 in the stroma. This represents a 10,000 fold concentration gradient for protons across the thylakoid membrane.

Source of Proton Gradient

The protons in the lumen come from three primary sources.

- Photolysis by photosystem II oxidises water to oxygen, protons and electrons in the lumen.

- The transfer of electrons from photosystem II to plastoquinone during non-cyclic electron transport consumes two protons from the stroma. These are released in the lumen when the reduced plastoquinol is oxidized by the cytochrome b6f protein complex on the lumen side of the thylakoid membrane. From the plastoquinone pool, electrons pass through the cytochrome b6f complex. This integral membrane assembly resembles cytochrome bc1.

- The reduction of plastoquinone by ferredoxin during cyclic electron transport also transfers two protons from the stroma to the lumen.

The proton gradient is also caused by the consumptions of protons in the stroma to make NADPH from NADP+ at the NADP reductase.

ATP Generation

The molecular mechanism of ATP (Adenosine triphosphate) generation in chloroplasts is similar to that in mitochondria and takes the required energy from the proton motive force (PMF). However, chloroplasts rely more on the chemical potential of the PMF to generate the potential energy required for ATP synthesis. The PMF is the sum of a proton chemical potential (given by the proton concentration gradient) and a trans-membrane electrical potential (given by charge separation across the membrane). Compared to the inner membranes of mitochondria, which have a significantly higher membrane potential due to charge separation, thylakoid membranes lack a charge gradient. To compensate for this, the 10,000 fold proton concentration gradient across the thylakoid membrane is much higher compared to a 10 fold gradient across the inner membrane of mitochondria. The resulting chemiosmotic potential between the lumen and stroma is high enough to drive ATP synthesis using the ATP synthase. As the protons travel back down the gradient through channels in ATP synthase, ADP + P_i are combined into ATP. In this manner, the light-dependent reactions are coupled to the synthesis of ATP via the proton gradient.

Thylakoid Membranes in Cyanobacteria

Cyanobacteria are photosynthetic prokaryotes with highly differentiated membrane systems. Cyanobacteria have an internal system of thylakoid membranes where the fully functional electron transfer chains of photosynthesis and respiration reside. The presence of different membrane systems lends these cells a unique complexity among bacteria. Cyanobacteria must be able to reorganize the membranes, synthesize new membrane lipids, and properly target proteins to the correct membrane system. The outer membrane, plasma membrane, and thylakoid membranes each have specialized roles in the cyanobacterial cell. Understanding the organization, functionality, protein composition and dynamics of the membrane systems remains a great challenge in cyanobacterial cell biology.

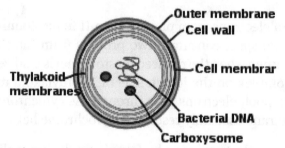

Thylakoids (green) inside a cyanobacterium (*Synechocystis*)

The thylakoid membranes of the cyanobacteria are not differentiated into granum and stroma regions as observed in plants. They form stacks of parallel sheets close to the cytoplasmic membrane with a low packing density. The relatively large distance between the thylakoids provides space for the external light harvesting antennae, the phycobilisomes. This macrostructure, as in the case of higher plants, shows some flexibility during changes in the physicochemical environment.

PI Curve

The PI (photosynthesis-irradiance) curve is a graphical representation of the empirical relationship between solar irradiance and photosynthesis. A derivation of the Michaelis–Menten curve, it shows the generally positive correlation between light intensity and photosynthetic rate. Plotted along the x-axis is the independent variable, light intensity (irradiance), while the y-axis is reserved for the dependent variable, photosynthetic rate.

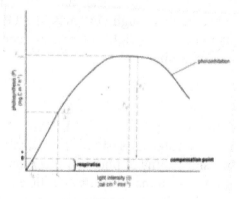

P v I curve

Introduction

The PI curve can be applied to terrestrial and marine reactions but is most commonly used to explain ocean-dwelling phytoplankton's photosynthetic response to changes in light intensity. Using this tool to approximate biological productivity is important because phytoplankton contribute ~50% of total global carbon fixation and are important suppliers to the marine food web.

Within the scientific community, the curve can be referred to as the PI, PE or Light Response Curve. While individual researchers may have their own preferences, all are readily acceptable for use in the literature. Regardless of nomenclature, the photosynthetic rate in question can be described in terms of carbon (C) fixed per unit per time. Since individuals vary in size, it is also useful to normalise C concentration to Chlorophyll a (an important photosynthetic pigment) to account for specific biomass.

History

As far back as 1905, marine researchers attempted to develop an equation to be used as the standard in establishing the relationship between solar irradiance and photosynthetic production. Several groups had relative success, but in 1976 a comparison study conducted by Alan Jassby and Trevor Platt, researchers at the Bedford Institute of Oceanography in Dartmouth, Nova Scotia, reached a conclusion that solidified the way in which a PI curve is developed. After evaluating the eight most-used equations, Jassby and Platt revealed that an adaptation of the Michaelis-Menten equation, previously used in enzyme kinetics, best served the relationship demonstration. Their findings were so conclusive that the Michaelis–Menten equation remains the standard for PI curve generation.

Equations

There are two simple derivations of the equation that are commonly used to generate the hyperbolic curve. The first assumes photosynthetic rate increases with increasing light intensity until Pmax is reached and continues to photosynthesise at the maximum rate thereafter.

$$P = P_{max}[I] / (KI + [I])$$

- P = photosynthetic rate at a given light intensity

  - Commonly denoted in units such as (mg C m-3 h-1) or (µg C µg Chl-a-1 h-1)

- Pmax = the maximum potential photosynthetic rate per individual

- [I] = a given light intensity

- o Commonly denoted in units such as (µMol photons m-2 s-1 or (Watts m-2 h-1)

- KI = half-saturation constant; the light intensity at which the photosyn- thetic rate proceeds at ½ Pmax

  - o Units reflect those used for [I]

Both Pmax and the initial slope of the curve, ΔP/ΔI, are species-specific, and are in- fluenced by a variety of factors, such as nutrient concentration, temperature and the physiological capabilities of the individual. Light intensity is influenced by latitudinal position and undergo daily and seasonal fluxes which will also affect the overall photo- synthetic capacity of the individual. These three parameters are predictable and can be used to predetermine the general PI curve a population should follow.

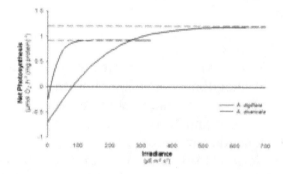

As can be seen in the graph, two species can have different responses to the same incre- mental changes in light intensity. Population A (in blue) has an initial rate higher than that of Population B (in red) and also exhibits a stronger rate change to increased light intensities at lower irradiance. Therefore, Population A will dominate in an environ- ment with lower light availability. Although Population B has a slower photosynthetic response to increases in light intensity its Pmax is higher than that of Population A. This allows for eventual population dominance at greater light intensities. There are many determining factors influencing population success; using the PI curve to elicit predictions of rate flux to environmental changes is useful for monitoring phytoplank- ton bloom dynamics and ecosystem stability.

The second equation accounts for the phenomenon of photoinhibition. In the upper few meters of the ocean, phytoplankton may be subjected to irradiance levels that dam- age the chlorophyll-a pigment inside the cell, subsequently decreasing photosynthetic rate. The response curve depicts photoinhibition as a decrease in photosynthetic rate at light intensities stronger than those necessary for achievement of Pmax.

$$P = P_{max}(1 - e^{-\alpha I / P_{max}})e^{-\beta I / P_{max}}$$

Terms not included in the above equation are:

- βI = light intensity at the start of photoinhibition

- αI = a given light intensity

Examples

Data sets showing interspecific differences and population dynamics.

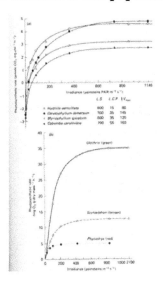

The hyperbolic response between photosynthesis and irradiance, depicted by the PI curve, is important for assessing phytoplankton population dynamics, which influence many aspects of the marine environment.

References

- Hall, David O.; Rao, Krishna (1999-06-24). Photosynthesis. Cambridge University Press. pp. 8–9. ISBN 9780521644976.

- Fitter, Alastair H.; Hay, Robert K. M. (2012-12-02). Environmental Physiology of Plants. Academic Press. p. 26. ISBN 9780080549811.

- Jones, Daniel (2003) [1917], Peter Roach, James Hartmann and Jane Setter, eds., English Pronouncing Dictionary, Cambridge: Cambridge University Press, ISBN 3-12-539683-2

- Alberts, Bruce (2002). Molecular biology of the cell (4. ed.). New York [u.a.]: Garland. ISBN 0-8153-4072-9.

- Sandelius, Anna Stina (2009). The Chloroplast: Interactions with the Environment. Springer. p. 18. ISBN 978-3-540-68696-5.

- Alberts, Bruce (2002). Molecular biology of the cell (4. ed.). New York [u.a.]: Garland. ISBN 0-8153-4072-9.

- Wise, R.R.; Hoober, J.K. (2007). The Structure and Function of Plastids. Springer. pp. 32–33. ISBN 978-1-4020-6570-5.

- J D, Rochaix (1998). The molecular biology of chloroplasts and mitochondria in Chlamydomonas. Dordrecht [u.a.]: Kluwer Acad. Publ. pp. 550–565. ISBN 978-0-7923-5174-0.

- Roberts, editor, Keith (2007). Handbook of plant science. Chichester, West Sussex, England: Wiley. p. 16. ISBN 978-0-470-05723-0.

- Campbell, Neil A.; Williamson, Brad; Heyden, Robin J. (2006). Biology: Exploring Life. Boston, Massachusetts: Pearson Prentice Hall. ISBN 978-0-13-250882-7.

- Burgess,, Jeremy (1989). An introduction to plant cell development. Cambridge: Cambridge university press. p. 56. ISBN 0-521-31611-1.

- Steer, Brian E.S. Gunning, Martin W. (1996). Plant cell biology : structure and function. Boston, Mass.: Jones and Bartlett Publishers. p. 20. ISBN 0-86720-504-0.

- Herrero A and Flores E (editor). (2008). The Cyanobacteria: Molecular Biology, Genomics and Evolution (1st ed.). Caister Academic Press. ISBN 1-904455-15-8. [1].

Organisms using Photosynthesis

The chapter studies organisms that utilize the process of photosynthesis to produce energy. These organisms include phototrophs, algae, cyanobacteria, purple bacteria, green sulfur bacteria and heliobacteria. The content studies their ecological distribution, taxonomy and classification. Photosynthesis is best understood in confluence with the major topics listed in the following chapter.

Phototroph

Terrestrial and aquatic phototrophs: plants grow on a fallen log floating in algae-rich water

Phototrophs are the organisms that carry out photon capture to acquire energy. They use the energy from light to carry out various cellular metabolic processes. It is a common misconception that phototrophs are obligatorily photosynthetic. Many, but not all, phototrophs often photosynthesize: they anabolically convert carbon dioxide into organic material to be utilized structurally, functionally, or as a source for later catabolic processes (e.g. in the form of starches, sugars and fats). All phototrophs either use electron transport chains or direct proton pumping to establish an electro-chemical gradient which is utilized by ATP synthase, to provide the molecular energy currency for the cell. Phototrophs can be either autotrophs or heterotrophs.

Photoautotroph

Most of the well-recognized phototrophs are autotrophic, also known as photoauto-trophs, and can fix carbon. They can be contrasted with chemotrophs that obtain their energy by the oxidation of electron donors in their environments. Photoautotrophs are capable of synthesizing their own food from inorganic substances using light as an energy source. Green plants and photosynthetic bacteria are photoautotrophs. Photoau-totrophic organisms are sometimes referred to as holophytic. Such organisms derive their energy for food synthesis from light and are capable of using carbon dioxide as their principal source of carbon.

Oxygenic photosynthetic organisms use chlorophyll for light-energy capture and oxi-dize water, "splitting" it into molecular oxygen. In contrast, anoxygenic photosynthetic bacteria have a substance called bacteriochlorophyll - which absorbs predominantly at non-optical wavelengths - for light-energy capture, live in aquatic environments, and will, using light, oxidize chemical substances such as hydrogen sulfide rather than wa-ter.

Ecology

In an ecological context, phototrophs are often the food source for neighboring het-erotrophic life. In terrestrial environments, plants are the predominant variety, while aquatic environments include a range of phototrophic organisms such as algae (e.g., kelp), other protists (such as euglena), phytoplankton, and bacteria (such as cyanobac-teria). The depth to which sunlight or artificial light can penetrate into water, so that photosynthesis may occur, is known as the photic zone.

Cyanobacteria, which are prokaryotic organisms which carry out oxygenic photosyn-thesis, occupy many environmental conditions, including fresh water, seas, soil, and li-chen. Cyanobacteria carry out plant-like photosynthesis because the organelle in plants that carries out photosynthesis is actually derived from an endosymbiosis cyanobacte-ria. This bacteria can use water as a source of electrons in order to perform CO_2 reduc-tion reactions. Evolutionarily, cyanobacteria's ability to survive in oxygenic conditions, which are considered toxic to most anaerobic bacteria, might have given the bacteria an adaptive advantage which could have allowed the cyanobacteria to populate more efficiently.

A *photolithoautotroph* is an autotrophic organism that uses light energy, and an in-organic electron donor (e.g., H_2O, H_2, H_2S), and CO_2 as its carbon source. Examples include plants.

Photoheterotroph

In contrast to photoautotrophs, photoheterotrophs are organisms that depend solely on light for their energy and principally on organic compounds for their carbon. Photoheterotrophs produce ATP through photophosphorylation but use environmentally obtained organic compounds to build structures and other bio-molecules.

Flowchart

- Autotroph
  - ○ Chemoautotroph
  - ○ Photoautotroph
- Heterotroph
  - ○ Chemoheterotroph
  - ○ Photoheterotroph

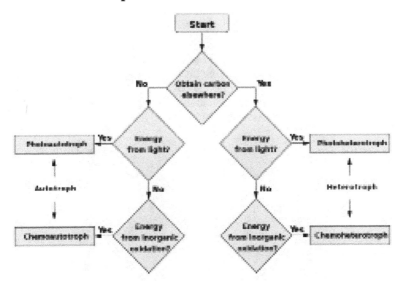

Flowchart to determine if a species is autotroph, heterotroph, or a subtype

Algae

Algae is an informal term for a large, diverse group of photosynthetic organisms which are not necessarily closely related and are thus polyphyletic. Included organisms range from unicellular genera, such as *Chlorella* and the diatoms, to multicellular forms, such as the giant kelp, a large brown alga which may grow up to 50 meters in length. Most are aquatic and autotrophic and lack many of the distinct cell and tissue types, such as stomata, xylem and phloem, which are found in land plants. The largest and most complex marine algae are called seaweeds, while the most complex freshwater forms are the Charophyta, a division of green algae which includes, for example, *Spirogyra* and the stoneworts.

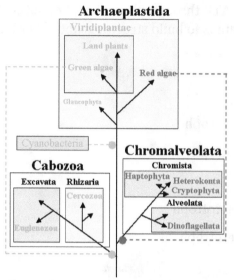

The lineage of algae according to Thomas Cavalier-Smith. The exact number and placement of endosymbiotic events is currently unknown, so this diagram can be taken only as a general guide. It represents the most parsimonious way of explaining the three types of endosymbiotic origins of plastids. These types include the endosymbiotic events of cyanobacteria, red algae and green algae, leading to the hypothesis of the supergroups Archaeplastida, Chromalveolata and Cabozoa respectively. Endosymbiotic events are noted by dotted lines.

A variety of algae growing on the sea bed in shallow waters

There is no generally accepted definition of algae. One definition is that algae "have chlorophyll as their primary photosynthetic pigment and lack a sterile covering of cells around their reproductive cells". Some authors exclude all prokaryotes and thus do not consider cyanobacteria (blue-green algae) as algae.

Algae constitute a polyphyletic group since they do not include a common ancestor, and although their plastids seem to have a single origin, from cyanobacteria, they were acquired in different ways. Green algae are examples of algae that have primary chloroplasts derived from endosymbiotic cyanobacteria. Diatoms and brown algae are examples of algae with secondary chloroplasts derived from an endosymbiotic red alga.

Algae exhibit a wide range of reproductive strategies, from simple asexual cell division to complex forms of sexual reproduction.

Algae lack the various structures that characterize land plants, such as the phyllids (leaf-like structures) of bryophytes, rhizoids in nonvascular plants, and the roots, leaves, and other organs that are found in tracheophytes (vascular plants). Most are phototrophic, although some are mixotrophic, deriving energy both from photosynthesis and uptake of organic carbon either by osmotrophy, myzotrophy, or phagotrophy. Some unicellular species of green algae, many golden algae, euglenids, dinoflagellates and other algae have become heterotrophs (also called colorless or apochlorotic algae), sometimes parasitic, relying entirely on external energy sources and have limited or no photosynthetic apparatus. Some other heterotrophic organisms, like the apicomplexans, are also derived from cells whose ancestors possessed plastids, but are not traditionally considered as algae. Algae have photosynthetic machinery ultimately derived from cyanobacteria that produce oxygen as a by-product of photosynthesis, unlike other photosynthetic bacteria such as purple and green sulfur bacteria. Fossilized filamentous algae from the Vindhya basin have been dated back to 1.6 to 1.7 billion years ago.

Etymology and Study

The singular *alga* is the Latin word for "seaweed" and retains that meaning in English. The etymology is obscure. Although some speculate that it is related to Latin *algēre*, "be cold", there is no known reason to associate seaweed with temperature. A more likely source is *alliga*, "binding, entwining."

The Latinization, *fūcus*, meant primarily the cosmetic rouge. The etymology is uncertain, but a strong candidate has long been some word related to the Biblical פוך (pūk), "paint" (if not that word itself), a cosmetic eye-shadow used by the ancient Egyptians and other inhabitants of the eastern Mediterranean. It could be any color: black, red, green, blue.

Accordingly, the modern study of marine and freshwater algae is called either phycology or algology, depending on whether the Greek or Latin root is used. The name *Fucus* appears in a number of taxa.

Classification

Most algae contain chloroplasts that are similar in structure to cyanobacteria. Chloroplasts contain circular DNA like that in cyanobacteria and presumably represent reduced endosymbiotic cyanobacteria. However, the exact origin of the chloroplasts is different among separate lineages of algae, reflecting their acquisition during different endosymbiotic events. The table below describes the composition of the three major groups of algae. Their lineage relationships are shown in the figure in the upper right.

Many of these groups contain some members that are no longer photosynthetic. Some retain plastids, but not chloroplasts, while others have lost plastids entirely.

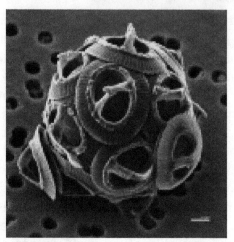

False-color Scanning electron micrograph of the unicellular coccolithophore *Gephyrocapsa oceanica*

Phylogeny based on plastid not nucleocytoplasmic genealogy:

Supergroup affiliation	Members	Endosymbiont	Summary
Primoplantae/ Archaeplastida	• Chlorophyta • Rhodophyta • Glaucophyta	Cyanobacteria	These algae have *primary* chloroplasts, i.e. the chloroplasts are surrounded by *two membranes* and probably developed through a single endosymbiotic event. The chloroplasts of red algae have chlorophylls *a* and *c* (often), and phycobilins, while those of green algae have chloroplasts with chlorophyll *a* and *b* without phycobilins. Land plants are pigmented similarly to green algae and probably developed from them, and thus Chlorophyta is a sister taxon to the plants; sometimes Chlorophyta, Charophyta and land plants are grouped together as Viridiplantae.
Excavata and Rhizaria	• Chlorarachnio-phytes • Euglenids	Green algae	These groups have green chloroplasts containing chlorophylls *a* and *b*. Their chloroplasts are surrounded by *four and three membranes* respectively, and were probably retained from ingested green algae. **Chlorarachniophytes**, which belong to the phylum Cercozoa, contain a small nucleomorph, which is a relict of the algae's nucleus. **Euglenids**, which belong to the phylum Euglenozoa, live primarily in freshwater and have chloroplasts with only three membranes. It has been suggested that the endosymbiotic green algae were acquired through myzocytosis rather than phagocytosis.

Chromista and Alveolata	• Heterokonts • Haptophyta • Cryptomonads • Dinoflagellates	Red algae	These groups have chloroplasts containing chlorophylls *a* and *c*, and phycobilins. The shape varies from plant to plant; they may be of discoid, plate-like, reticulate, cup-shaped, spiral or ribbon shaped. They have one or more pyrenoids to preserve protein and starch. The latter chlorophyll type is not known from any prokaryotes or primary chloroplasts, but genetic similarities with red algae suggest a relationship there.
			In the first three of these groups (**Chromista**), the chloroplast has four membranes, retaining a nucleomorph in Cryptomonads, and they likely share a common pigmented ancestor, although other evidence casts doubt on whether the Heterokonts, Haptophyta, and Cryptomonads are in fact more closely related to each other than to other groups.
			The typical **dinoflagellate** chloroplast has three membranes, but there is considerable diversity in chloroplasts within the group, and it appears that there were a number of endosymbiotic events. The Apicomplexa, a group of closely related parasites, also have plastids called apicoplasts. Apicoplasts are not photosynthetic, but appear to have a common origin with Dinoflagellate chloroplasts.

HISTORIA

FVCORVM

AVCTORE

SAMVEL GOTTLIEB GMELIN,

MED. DOCT. ACADEM. IMPER. PETROPOL. BOTA-
NICES PROFESSORE ET MEMBRO ORDINARIO

PETROPOLI
EX TYPOGRAPHIA ACADEMIAE SCIENTIARVM
cIɔ Iɔcc LXVIII.

Title page of Gmelin's *Historia Fucorum*, dated 1768

Linnaeus, in *Species Plantarum* (1753), the starting point for modern botanical nomenclature, recognized 14 genera of algae, of which only 4 are currently considered among algae. In *Systema Naturae*, Linnaeus described the genera *Volvox*, *Corallina* and a species of *Acetabularia* (as *Madrepora*), among the animals.

In 1768, Samuel Gottlieb Gmelin (1744–1774) published the *Historia Fucorum*, the first work dedicated to marine algae and the first book on marine biology to use the then new binomial nomenclature of Linnaeus. It included elaborate illustrations of seaweed and marine algae on folded leaves.

W.H.Harvey (1811–1866) and Lamouroux (1813) were the first to divide macroscopic algae into four divisions based on their pigmentation. This is the first use of a biochemical criterion in plant systematics. Harvey's four divisions are: red algae (Rhodospermae), brown algae (Melanospermae), green algae (Chlorospermae) and Diatomaceae.

At this time, microscopic algae were discovered and reported by a different group of workers (e.g., O. F. Müller and Ehrenberg) studying the Infusoria (microscopic organisms). Unlike macroalgae, which were clearly viewed as plants, microalgae were frequently considered animals because they are often motile. Even the non-motile (coccoid) microalgae were sometimes merely seen as stages of the life cycle of plants, macroalgae or animals.

Although used as a taxonomic category in some pre-Darwinian classifications, e.g., Linnaeus (1753), de Jussieu (1789), Horaninow (1843), Agassiz (1859), Wilson & Cassin (1864), in further classifications, the "algae" are seen as an artificial, polyphyletic group.

Throughout the 20th century, most classifications treated the following groups as divisions or classes of algae: cyanophytes, rhodophytes, chrysophytes, xanthophytes, bacillariophytes, phaeophytes, pyrrhophytes (cryptophytes and dinophytes), euglenophytes and chlorophytes. Later, many new groups were discovered (e.g., Bolidophyceae), and others were splintered from older groups: charophytes and glaucophytes (from chlorophytes), many heterokontophytes (e.g., synurophytes from chrysophytes, or eustigmatophytes from xanthophytes), haptophytes (from chrysophytes) and chlorarachniophytes (from xanthophytes).

With the abandonment of plant-animal dichotomous classification, most groups of algae (sometimes all) were included in Protista, later also abandoned in favour of Eukaryota. However, as a legacy of the older plant life scheme, some groups that were also treated as protozoans in the past still have duplicated classifications.

Some parasitic algae (e.g., the green algae *Prototheca* and *Helicosporidium*, parasites of metazoans, or *Cephaleuros*, parasites of plants) were originally classified as fungi, sporozoans or protistans of incertae sedis, while others (e.g., the green algae *Phyllosiphon* and *Rhodochytrium*, parasites of plants, or the red algae *Pterocladiophila* and *Gelidiocolax mammillatus*, parasites of other red algae, or the dinoflagellates *Oodinium*, parasites of fish) had their relationship with algae conjectured early. In other cases, some groups were originally characterized as parasitic algae (e.g., *Chlorochytrium*), but later were seen as endophytic algae. Furthermore, groups like the apicomplexans are also parasites derived from ancestors that possessed plastids, but are not included in any group traditionally seen as algae.

Relationship to Land Plants

The first land plants probably evolved from shallow freshwater charophyte algae much like *Chara* almost 500 million years ago. These probably had an isomorphic alternation of generations and were probably filamentous. Fossils of isolated land plant spores suggest land plants may have been around as long as 475 million years ago.

Morphology

The kelp forest exhibit at the Monterey Bay Aquarium. A three-dimensional, multicellular thallus

A range of algal morphologies are exhibited, and convergence of features in unrelated groups is common. The only groups to exhibit three-dimensional multicellular thalli are the reds and browns, and some chlorophytes. Apical growth is constrained to subsets of these groups: the florideophyte reds, various browns, and the charophytes. The form of charophytes is quite different from those of reds and browns, because they have distinct nodes, separated by internode 'stems'; whorls of branches reminiscent of the horsetails occur at the nodes. Conceptacles are another polyphyletic trait; they appear in the coralline algae and the Hildenbrandiales, as well as the browns.

Most of the simpler algae are unicellular flagellates or amoeboids, but colonial and non-motile forms have developed independently among several of the groups. Some of the more common organizational levels, more than one of which may occur in the life cycle of a species, are

- *Colonial*: small, regular groups of motile cells

- *Capsoid*: individual non-motile cells embedded in mucilage

- *Coccoid*: individual non-motile cells with cell walls

- *Palmelloid*: non-motile cells embedded in mucilage

- *Filamentous*: a string of non-motile cells connected together, sometimes

branching

- *Parenchymatous*: cells forming a thallus with partial differentiation of tissues

In three lines, even higher levels of organization have been reached, with full tissue differentiation. These are the brown algae,—some of which may reach 50 m in length (kelps)—the red algae, and the green algae. The most complex forms are found among the green algae, in a lineage that eventually led to the higher land plants. The point where these non-algal plants begin and algae stop is usually taken to be the presence of reproductive organs with protective cell layers, a characteristic not found in the other alga groups.

Physiology

Many algae, particularly members of the Characeae, have served as model experimental organisms to understand the mechanisms of the water permeability of membranes, osmoregulation, turgor regulation, salt tolerance, cytoplasmic streaming, and the generation of action potentials.

Phytohormones are found not only in higher plants, but in algae too.

Symbiotic Algae

Some species of algae form symbiotic relationships with other organisms. In these symbioses, the algae supply photosynthates (organic substances) to the host organism providing protection to the algal cells. The host organism derives some or all of its energy requirements from the algae. Examples are as follows.

Lichens

Rock lichens in Ireland

Lichens are defined by the International Association for Lichenology to be "an association of a fungus and a photosynthetic symbiont resulting in a stable vegetative body having a specific structure." The fungi, or mycobionts, are mainly from the Ascomycota with a few from the Basidiomycota. They are not found alone in nature; but when they began to associate is not known. One mycobiont associates with the same phycobiont species, rarely two, from the green algae, except that alternatively the mycobiont may associate with a species of cyanobacteria (hence "photobiont" is the more accurate term). A photobiont may be associated with many different mycobionts or may live independently; accordingly, lichens are named and classified as fungal species. The association is termed a morphogenesis because the lichen has a form and capabilities not possessed by the symbiont species alone (they can be experimentally isolated). It is possible that the photobiont triggers otherwise latent genes in the mycobiont.

Coral Reefs

Floridian coral reef

Coral reefs are accumulated from the calcareous exoskeletons of marine invertebrates of the order Scleractinia (stony corals). These animals metabolize sugar and oxygen to obtain energy for their cell-building processes, including secretion of the exoskeleton, with water and carbon dioxide as byproducts. Dinoflagellates (algal protists) are often endosymbionts in the cells of the coral-forming marine invertebrates, where they accelerate host-cell metabolism by generating immediately available sugar and oxygen through photosynthesis using incident light and the carbon dioxide produced by the host. Reef-building stony corals (hermatypic corals) require endosymbiotic algae from the genus *Symbiodinium* to be in a healthy condition. The loss of *Symbiodinium* from the host is known as coral bleaching, a condition which leads to the deterioration of a reef.

Sea Sponges

Green algae live close to the surface of some sponges, for example, breadcrumb sponge (*Halichondria panicea*). The alga is thus protected from predators; the sponge is provided with oxygen and sugars which can account for 50 to 80% of sponge growth in some species.

Life-cycle

Rhodophyta, Chlorophyta and Heterokontophyta, the three main algal divisions, have life-cycles which show considerable variation and complexity. In general, there is an asexual phase where the seaweed's cells are diploid, a sexual phase where the cells are haploid followed by fusion of the male and female gametes. Asexual reproduction permits efficient population increases, but less variation is possible. Commonly, in sexual reproduction of unicellular and colonial algae, two specialized sexually compatible haploid gametes make physical contact and fuse to form a zygote. To ensure a successful mating, the development and release of gametes is highly synchronized and regulated; pheromones may play a key role in these processes. Sexual reproduction allows for more variation and provides the benefit of efficient recombinational repair of DNA damages during meiosis, a key stage of the sexual cycle. However, sexual reproduction is more costly than asexual reproduction. Meiosis has been shown to occur in many different species of algae.

Numbers

Algae on coastal rocks at Shihtiping in Taiwan

The *Algal Collection of the US National Herbarium* (located in the National Museum of Natural History) consists of approximately 320,500 dried specimens, which, although not exhaustive (no exhaustive collection exists), gives an idea of the order of magnitude of the number of algal species (that number remains unknown). Estimates vary widely. For example, according to one standard textbook, in the British Isles the *UK Biodiversity Steering Group Report* estimated there to be 20000 algal species in the UK. Another checklist reports only about 5000 species. Regarding the difference of about 15000 species, the text concludes: "It will require many detailed field surveys before it is possible to provide a reliable estimate of the total number of species ..."

Regional and group estimates have been made as well:

- 5000–5500 species of red algae worldwide

- "some 1300 in Australian Seas"

- 400 seaweed species for the western coastline of South Africa, and 212 species from the coast of KwaZulu-Natal. Some of these are duplicates, as the range extends across both coasts, and the total recorded is probably about 500 species. Most of these are listed in List of seaweeds of South Africa. These exclude phytoplankton and crustose corallines.

- 669 marine species from California (US)

- 642 in the check-list of Britain and Ireland

and so on, but lacking any scientific basis or reliable sources, these numbers have no more credibility than the British ones mentioned above. Most estimates also omit microscopic algae, such as phytoplankton.

The most recent estimate suggests 72,500 algal species worldwide.

Distribution

The distribution of algal species has been fairly well studied since the founding of phytogeography in the mid-19th century AD. Algae spread mainly by the dispersal of spores analogously to the dispersal of Plantae by seeds and spores. This dispersal can be accomplished by air, water, or other organisms. Due to this, spores can be found in a variety of environments: fresh and marine waters, air, soil, and in or on other organisms. Whether a spore is to grow into an organism depends on the combination of the species and the environmental conditions of where the spore lands.

The spores of fresh-water algae are dispersed mainly by running water and wind, as well as by living carriers. However, not all bodies of water can carry all species of algae, as the chemical composition of certain water bodies will limit the algae that can survive within it. Marine spores are often spread by ocean currents. Ocean water presents many vastly-different habitats based on temperature and nutrient-availability, resulting in phytogeographic zones, regions and provinces.

To some degree, the distribution of algae is subject to floristic discontinuities caused by geographical features, such as Antarctica, long distances of ocean or general land masses. It is therefore possible to identify species occurring by locality, such as "Pacific Algae" or "North Sea Algae". When they occur out of their localities, it is usually possible to hypothesize a transport mechanism, such as the hulls of ships. For example, *Ulva reticulata* and *Ulva fasciata* travelled from the mainland to Hawaii in this manner.

Mapping is possible for select species only: "there are many valid examples of confined distribution patterns." For example, *Clathromorphum* is an arctic genus and is not mapped far south of there. On the other hand, scientists regard the overall data as insufficient due to the "difficulties of undertaking such studies."

Ecology

Algae are prominent in bodies of water, common in terrestrial environments and are found in unusual environments, such as on snow and on ice. Seaweeds grow mostly in shallow marine waters, under 100 metres (330 ft); however, some have been recorded to a depth of 360 metres (1,180 ft).

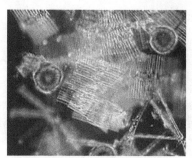

Phytoplankton, Lake Chuzenji

The various sorts of algae play significant roles in aquatic ecology. Microscopic forms that live suspended in the water column (phytoplankton) provide the food base for most marine food chains. In very high densities (algal blooms) these algae may discolor the water and outcompete, poison, or asphyxiate other life forms.

Algae can be used as indicator organisms to monitor pollution in various aquatic systems. In many cases, algal metabolism is sensitive to various pollutants. Due to this, the species composition of algal populations may shift in the presence of chemical pollutants. To detect these changes, algae can be sampled from the environment and maintained in laboratories with relative ease.

On the basis of their habitat, algae can be categorized as: aquatic (planktonic, benthic, marine, freshwater), terrestrial, aerial (subareial), lithophytic, halophytic (or euryhaline), psammon, thermophilic, cryophilic, epibiont (epiphytic, epizoic), endosymbiont (endophytic, endozoic), parasitic, calcifilic or lichenic (phycobiont).

Cultural Associations

In Classical Chinese, the word 藻 is used both for "algae" and (in the modest tradition of the imperial scholars) for "literary talent". The third island in Kunming Lake beside the Summer Palace in Beijing is known as the Zaojian Tang Dao which thus simultaneously means "Island of the Algae-Viewing Hall" and "Island of the Hall for Reflecting on Literary Talent".

Agar

Agar, a gelatinous substance derived from red algae, has a number of commercial uses. It is a good medium on which to grow bacteria and fungi as most microorganisms cannot digest agar.

Alginates

Alginic acid, or alginate, is extracted from brown algae. Its uses range from gelling agents in food, to medical dressings. Alginic acid also has been used in the field of biotechnology as a biocompatible medium for cell encapsulation and cell immobilization.

Molecular cuisine is also a user of the substance for its gelling properties, by which it becomes a delivery vehicle for flavours.

Between 100,000 and 170,000 wet tons of *Macrocystis* are harvested annually in New Mexico for alginate extraction and abalone feed.

Energy Source

To be competitive and independent from fluctuating support from (local) policy on the long run, biofuels should equal or beat the cost level of fossil fuels. Here, algae based fuels hold great promise, directly related to the potential to produce more biomass per unit area in a year than any other form of biomass. The break-even point for algae-based biofuels is estimated to occur by 2025.

Fertilizer

Seaweed-fertilized gardens on Inisheer

For centuries, seaweed has been used as a fertilizer; George Owen of Henllys writing in the 16th century referring to drift weed in South Wales:

This kind of ore they often gather and lay on great heapes, where it heteth and rotteth, and will have a strong and loathsome smell; when being so rotten they cast on the land, as they do their muck, and thereof springeth good corn, especially barley ... After spring-tydes or great rigs of the sea, they fetch it in sacks on horse backes, and carie the same three, four, or five miles, and cast it on the lande, which doth very much better the ground for corn and grass.

Today, algae are used by humans in many ways; for example, as fertilizers, soil conditioners and livestock feed. Aquatic and microscopic species are cultured in clear tanks or ponds and are either harvested or used to treat effluents pumped through the ponds. Algaculture on a large scale is an important type of aquaculture in some places. Maerl is commonly used as a soil conditioner.

Nutrition

Dulse, a type of food

Naturally growing seaweeds are an important source of food, especially in Asia. They provide many vitamins including: A, B_1, B_2, B_6, niacin and C, and are rich in iodine, potassium, iron, magnesium and calcium. In addition commercially cultivated microalgae, including both algae and cyanobacteria, are marketed as nutritional supplements, such as Spirulina, Chlorella and the Vitamin-C supplement, Dunaliella, high in beta-carotene.

Algae are national foods of many nations: China consumes more than 70 species, including *fat choy*, a cyanobacterium considered a vegetable; Japan, over 20 species; Ireland, dulse; Chile, cochayuyo. Laver is used to make "laver bread" in Wales where it is known as *bara lawr*; in Korea, gim; in Japan, nori and aonori. It is also used along the west coast of North America from California to British Columbia, in Hawaii and by the Māori of New Zealand. Sea lettuce and badderlocks are a salad ingredient in Scotland, Ireland, Greenland and Iceland.

The oils from some algae have high levels of unsaturated fatty acids. For example, *Parietochloris incisa* is very high in arachidonic acid, where it reaches up to 47% of the triglyceride pool. Some varieties of algae favored by vegetarianism and veganism contain the long-chain, essential omega-3 fatty acids, docosahexaenoic acid (DHA) and eicosapentaenoic acid (EPA). Fish oil contains the omega-3 fatty acids, but the original source is algae (microalgae in particular), which are eaten by marine life such as copepods and are passed up the food chain. Algae have emerged in recent years as a popular source of omega-3 fatty acids for vegetarians who cannot get long-chain EPA and DHA from other vegetarian sources such as flaxseed oil, which only contains the short-chain alpha-linolenic acid (ALA).

Pollution Control

- Sewage can be treated with algae, reducing the usage of large amounts of toxic chemicals that would otherwise be needed.

- Algae can be used to capture fertilizers in runoff from farms. When subsequently harvested, the enriched algae itself can be used as fertilizer.

- Aquariums and ponds can be filtered using algae, which absorb nutrients from the water in a device called an algae scrubber, also known as an algae turf scrubber (A T S) .

Agricultural Research Service scientists found that 60–90% of nitrogen runoff and 70–100% of phosphorus runoff can be captured from manure effluents using a horizontal algae scrubber, also called an algal turf scrubber (ATS). Scientists developed the ATS, which consists of shallow, 100-foot raceways of nylon netting where algae colonies can form, and studied its efficacy for three years. They found that algae can readily be used to reduce the nutrient runoff from agricultural fields and increase the quality of water flowing into rivers, streams, and oceans. Researchers collected and dried the nutrient-rich algae from the ATS and studied its potential as an organic fertilizer. They found that cucumber and corn seedlings grew just as well using ATS organic fertilizer as they did with commercial fertilizers. Algae scrubbers, using bubbling upflow or vertical waterfall versions, are now also being used to filter aquariums and ponds.

Bioremediation

The alga *Stichococcus bacillaris*, has been seen to colonize silicone resins used at archaeological sites; biodegrading the synthetic substance.

Pigments

The natural pigments (carotenoids and chlorophylls) produced by algae can be used as an alternative to chemical dyes and coloring agents. The presence of some individual alga pigments, together with specific pigment concentrations ratios, are taxon-specific: analysis of their concentrations with various analytical methods, particularly high-performance liquid chromatography (HPLC), can therefore offer deep insight into the taxonomic composition and relative abundance of natural alga populations in sea water samples.

Stabilizing Substances

Carrageenan, from the red alga *Chondrus crispus*, is used as a stabilizer in milk products.

Cyanobacteria

Like other prokaryotes, cyanobacteria have no internal membrane bound organelles. They perform photosynthesis in distinctive folds in the outer membrane, unlike green plants which use organelles called chloroplasts. Symbiogenesis argues that the chloroplasts found in plants and eukaryotic algae evolved from cyanobacterial ancestors via endosymbiosis.

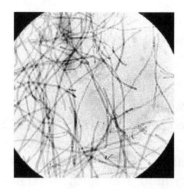

Cyanobacteria is a phylum of bacteria that obtain their energy through photosynthesis. The name "cyanobacteria" comes from the color of the bacteria. They are often called blue-green algae, although the name is sometimes considered a misnomer because cyanobacteria are prokaryotes and the term "algae" is often reserved for eukaryotes.

By producing oxygen as a byproduct of photosynthesis, cyanobacteria are thought to have converted the early reducing atmosphere into an oxidizing one, causing the "rusting of the Earth" and causing the Great Oxygenation Event, dramatically changing the composition of life forms on Earth by stimulating biodiversity and leading to the near-extinction of anaerobic organisms (that is, oxygen-intolerant).

Cyanobacteria are a group of photosynthetic, nitrogen fixing bacteria that live in a wide variety of habitats such as moist soils and in water. They may be free-living or form symbiotic relationships with plants or with lichen-forming fungi as in the lichen genus *Peltigera*. They range from unicellular to filamentous and include colonial species. Colonies may form filaments, sheets, or even hollow balls. Some filamentous species can differentiate into several different cell types: vegetative cells, the normal, photosynthetic cells that are formed under favorable growing conditions; akinetes, climate-resistant spores that may form when environmental conditions become harsh; and thick-walled heterocysts, which contain the enzyme nitrogenase, vital for nitrogen fixation.

Nitrogen Fixation

Cyanobacteria can fix atmospheric nitrogen in anaerobic conditions by means of specialized cells called heterocysts. Heterocysts may also form under the appropriate environmental conditions (anoxic) when fixed nitrogen is scarce. Heterocyst-forming species are specialized for nitrogen fixation and are able to fix nitrogen gas into ammonia (NH_3), nitrites ($NO-2$) or nitrates ($NO-3$), which can be absorbed by plants and converted to protein and nucleic acids (atmospheric nitrogen is not bioavailable to plants, except for those having endosymbiotic nitrogen-fixing bacteria, especially the Fabaceae family, among others). Most importantly they are unicellular.

Free-living cyanobacteria are present in the water column in rice paddies, and cyanobacteria can be found growing as epiphytes on the surfaces of the green alga, Chara, where they may fix nitrogen. Cyanobacteria such as (*Anabaena*, a symbiont of the aquatic fern *Azolla*), can provide rice plantations with biofertilizer.

Morphology

Colonies of Nostoc pruniforme

Many cyanobacteria form motile filaments of cells, called hormogonia, that travel away from the main biomass to bud and form new colonies elsewhere. The cells in a hormogonium are often thinner than in the vegetative state, and the cells on either end of the motile chain may be tapered. To break away from the parent colony, a hormogonium often must tear apart a weaker cell in a filament, called a necridium.

Each individual cell of a cyanobacterium typically has a thick, gelatinous cell wall. They lack flagella, but hormogonia of some species can move about by gliding along surfaces. Many of the multicellular filamentous forms of *Oscillatoria* are capable of a waving motion; the filament oscillates back and forth. In water columns, some cyanobacteria float by forming gas vesicles, as in archaea. These vesicles are not organelles as such. They are not bounded by lipid membranes, but by a protein sheath.

Ecology

Cyanobacterial bloom near Fiji

Cyanobacteria can be found in almost every terrestrial and aquatic habitat—oceans, fresh water, damp soil, temporarily moistened rocks in deserts, bare rock and soil, and even Antarctic rocks. They can occur as planktonic cells or form phototrophic biofilms. They are found in almost every endolithic ecosystem. A few are endosymbionts in lichens, plants, various protists, or sponges and provide energy for the host. Some live in the fur of sloths, providing a form of camouflage.

Aquatic cyanobacteria are known for their extensive and highly visible blooms that can form in both freshwater and marine environments. The blooms can have the appearance of blue-green paint or scum. These blooms can be toxic, and frequently lead to the closure of recreational waters when spotted. Marine bacteriophages are significant parasites of unicellular marine cyanobacteria.

Some of these organisms contribute significantly to global ecology and the oxygen cycle. The tiny marine cyanobacterium *Prochlorococcus* was discovered in 1986 and accounts for more than half of the photosynthesis of the open ocean. Many cyanobacteria even display the circadian rhythms that were once thought to exist only in eukaryotic cells.

"Cyanobacteria are arguably the most successful group of microorganisms on earth. They are the most genetically diverse; they occupy a broad range of habitats across all latitudes, widespread in freshwater, marine, and terrestrial ecosystems, and they are found in the most extreme niches such as hot springs, salt works, and hypersaline bays. Photoautotrophic, oxygen-producing cyanobacteria created the conditions in the planet's early atmosphere that directed the evolution of aerobic metabolism and eukaryotic photosynthesis. Cyanobacteria fulfill vital ecological functions in the world's oceans, being important contributors to global carbon and nitrogen budgets." – Stewart and Falconer

Photosynthesis

While contemporary cyanobacteria are linked to the plant kingdom as descendants of the endosymbiotic progenitor of the chloroplast, there are several features which are unique to this group.

Carbon Fixation

Cyanobacteria use the energy of sunlight to drive photosynthesis, a process where the energy of light is used to split water molecules into oxygen, protons, and electrons. Because they are aquatic organisms, they typically employ several strategies which are collectively known as a "carbon concentrating mechanism" to aid in the acquisition of inorganic carbon (CO_2 or bicarbonate). Among the more specific strategies is the widespread prevalence of the bacterial microcompartments known as carboxysomes. These icosahedral structures are composed of hexameric shell proteins that assemble into cage-like structures that can be several hundreds of nanometers in diameter. It is believed that these structures tether the CO_2-fixing enzyme, RuBisCO, to the interior of the shell, as well as the enzyme carbonic anhydrase, using the paradigm of metabolic channeling to enhance the local CO_2 concentrations and thus increase the efficiency of the RuBisCO enzyme.

Electron Transport

In contrast to chloroplast-containing eukaryotes, cyanobacteria lack compartmentalization of their thylakoid membranes, which are contiguous with the plasma membrane. Thus, the protein complexes involved in respiratory energy metabolism share several mobile energy carrier pools (e.g., the Quinone pool, cytochrome c, ferredoxins), so photosynthetic and respiratory metabolism interact with each other. Furthermore, there is a tremendous diversity among the respiratory components between species. Thus cyanobacteria can be said to have a "branched electron transport chain", analogous to the situation in purple bacteria.

While most of the high-energy electrons derived from water are used by the cyanobacterial cells for their own needs, a fraction of these electrons may be donated to the external environment via electrogenic activity.

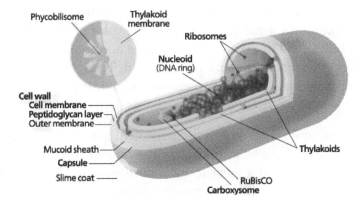

Metabolism and Organelles

As prokaryotes, cyanobacteria do not have nuclei or an internal membrane system. In most forms, the photosynthetic machinery is embedded into folds of the external cell membrane, called thylakoids. Cyanobacteria get their colour from the bluish pigment phycocyanin, which they use to capture light for photosynthesis. In general, photosynthesis in cyanobacteria uses water as an electron donor and produces oxygen as a byproduct, though some may also use hydrogen sulfide a process which occurs among other photosynthetic bacteria such as the purple sulfur bacteria. Carbon dioxide is reduced to form carbohydrates via the Calvin cycle.The large amounts of oxygen in the atmosphere are considered to have been first created by the activities of ancient cyanobacteria. They are often found as symbionts with a number of other groups of organisms such as fungi (lichens), corals, pteridophytes (*Azolla*), angiosperms (*Gunnera*), etc.

Many cyanobacteria are able to reduce nitrogen and carbon dioxide under aerobic conditions, a fact that may be responsible for their evolutionary and ecological success. The water-oxidizing photosynthesis is accomplished by coupling the activity of photosystem (PS) II and I (Z-scheme). In anaerobic conditions, they are able to use only PS I—cyclic photophosphorylation—with electron donors other than water (hydrogen sulfide, thiosulphate, or even molecular hydrogen) just like purple photosynthetic bacteria. Furthermore, they share an archaeal property, the ability to reduce elemental sulfur by anaerobic respiration in the dark. Their photosynthetic electron transport shares the same compartment as the components of respiratory electron transport. Their plasma membrane contains only components of the respiratory chain, while the thylakoid membrane hosts an interlinked respiratory and photosynthetic electron transport chain. The terminal oxidases in the thylakoid membrane respiratory/photosynthetic electron transport chain are essential for survival to rapid light changes, although not for dark maintenance under conditions where cells are not light stressed.

Attached to the thylakoid membrane, phycobilisomes act as light-harvesting antennae for the photosystems. The phycobilisome components (phycobiliproteins) are responsible for the blue-green pigmentation of most cyanobacteria. The variations on this theme are due mainly to carotenoids and phycoerythrins that give the cells their red-brownish coloration. In some cyanobacteria, the color of light influences the composition of phycobilisomes. In green light, the cells accumulate more phycoerythrin, whereas in red light they produce more phycocyanin. Thus, the bacteria appear green in red light and red in green light. This process of complementary chromatic adaptation is a way for the cells to maximize the use of available light for photosynthesis.

A few genera lack phycobilisomes and have chlorophyll b instead (*Prochloron, Prochlorococcus, Prochlorothrix*). These were originally grouped together as the prochlorophytes or chloroxybacteria, but appear to have developed in several different lines of cyanobacteria. For this reason, they are now considered as part of the cyanobacterial group.

There are some groups capable of heterotrophic growth, while others are parasitic, causing diseases in invertebrates or eukaryotic algae (e.g., the black band disease).

Relationship to Chloroplasts

Chloroplasts found in eukaryotes (algae and plants) appear to have evolved from an endosymbiotic relation with cyanobacteria. This endosymbiotic theory is supported by various structural and genetic similarities. Primary chloroplasts are found among the "true plants" or green plants – species ranging from sea lettuce to evergreens and flowers that contain chlorophyll *b* – as well as among the red algae and glaucophytes, marine species that contain phycobilins. It now appears that these chloroplasts probably had a single origin, in an ancestor of the clade called Archaeplastida. However this does not necessitate origin from cyanobacteria themselves; microbiology is still undergoing profound classification changes and entire domains (such as Archaea) are poorly mapped and understood. Other algae likely took their chloroplasts from these forms by secondary endosymbiosis or ingestion.

Classification

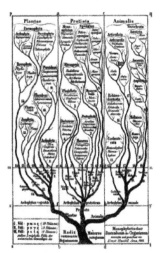

Tree of Life in Generelle Morphologie der Organismen (1866). Note the location of the genus Nostoc with algae and not with bacteria (kingdom "Monera")

Historically, bacteria were first classified as plants constituting the class Schizomycetes, which along with the Schizophyceae (blue-green algae/Cyanobacteria) formed the phylum Schizophyta, then in the phylum Monera in the kingdom Protista by Haeckel in 1866, comprising *Protogens, Protamaeba, Vampyrella, Protomonae,* and *Vibrio*, but not *Nostoc* and other cyanobacteria, which were classified with algae, later reclassified as the *Prokaryotes* by Chatton.

The cyanobacteria were traditionally classified by morphology into five sections, referred to by the numerals I-V. The first three – Chroococcales, Pleurocapsales, and Oscillatoriales – are not supported by phylogenetic studies. The latter two – Nostocales and Stigonematales – are monophyletic, and make up the heterocystous cyanobacteria.

The members of Chroococales are unicellular and usually aggregate in colonies. The classic taxonomic criterion has been the cell morphology and the plane of cell division. In Pleurocapsales, the cells have the ability to form internal spores (baeocytes). The rest of the sections include filamentous species. In Oscillatoriales, the cells are uniseriately arranged and do not form specialized cells (akinetes and heterocysts). In Nostocales and Stigonematales, the cells have the ability to develop heterocysts in certain conditions. Stigonematales, unlike Nostocales, include species with truly branched trichomes.

Most taxa included in the phylum or division Cyanobacteria have not yet been validly published under the Bacteriological Code, except:

- The classes Chroobacteria, Hormogoneae, and Gloeobacteria

- The orders Chroococcales, Gloeobacterales, Nostocales, Oscillatoriales, Pleuro-capsales, and Stigonematales

- The families Prochloraceae and Prochlorotrichaceae

- The genera *Halospirulina, Planktothricoides, Prochlorococcus, Prochloron,* and *Prochlorothrix*

The remainder are validly published under the International Code of Nomenclature for algae, fungi, and plants.

Formerly, some bacteria, like *Beggiatoa*, were thought to be colorless Cyanobacteria.

Earth History

Stromatolites are layered biochemical accretionary structures formed in shallow water by the trapping, binding, and cementation of sedimentary grains by biofilms (microbial mats) of microorganisms, especially cyanobacteria.

Stromatolites left behind by cyanobacteria are the oldest known fossils of life on Earth. This one-billion-year-old fossil is from Glacier National Park in Montana.

During the Precambrian, stromatolite communities of microorganisms grew in most marine and non-marine environments in the photic zone. After the Cambrian explosion of marine animals, grazing on the stromatolite mats by herbivores greatly reduced the occurrence of the stromatolites in marine environments. Since then, they are found mostly in hypersaline conditions where grazing invertebrates can't live (e.g. Shark Bay, Western Australia). Stromatolites provide ancient records of life on Earth by fossil remains which might date from more than 3.5 Ga ago, but this is disputed. As of 2010 the oldest undisputed evidence of cyanobacteria is from 2.1 Ga ago, but there is some evidence for them as far back as 2.7 Ga ago. Oxygen levels in the atmosphere remained around or below 1% of today's level until 2.4 Ga ago (the Great Oxygenation Event). The rise in oxygen may have caused a fall in methane levels, and triggered the Huronian glaciation from around 2.4 to 2.1 Ga ago. In this way, cyanobacteria may have killed off much of the other bacteria of the time.

Oncolites are sedimentary structures composed of oncoids, which are layered structures formed by cyanobacterial growth. Oncolites are similar to stromatolites, but instead of forming columns, they form approximately spherical structures that were not attached to the underlying substrate as they formed. The oncoids often form around a central nucleus, such as a shell fragment, and a calcium carbonate structure is deposited by encrusting microbes. Oncolites are indicators of warm waters in the photic zone, but are also known in contemporary freshwater environments. These structures rarely exceed 10 cm in diameter.

Biotechnology and Applications

The unicellular cyanobacterium *Synechocystis* sp. PCC6803 was the third prokaryote and first photosynthetic organism whose genome was completely sequenced. It continues to be an important model organism. *Cyanothece* ATCC 51142 is an important diazotrophic model organism. The smallest genomes have been found in *Prochlorococcus* spp. (1.7 Mb) and the largest in *Nostoc punctiforme* (9 Mb). Those of *Calothrix* spp. are estimated at 12–15 Mb, as large as yeast.

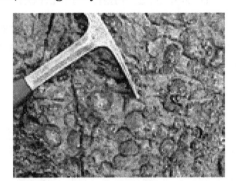

Oncolites from the Late Devonian Alamo bolide impact in Nevada

Recent research has suggested the potential application of cyanobacteria to the generation of renewable energy by converting sunlight into electricity. Internal photosynthetic pathways can be coupled to chemical mediators that transfer electrons to external electrodes. Currently, efforts are underway to commercialize algae-based fuels such as diesel, gasoline, and jet fuel.

Cyanobacteria cultured in specific media: Cyanobacteria can be helpful in agriculture as they have the ability to fix atmospheric nitrogen in soil.

Researchers from a company called Algenol have cultured genetically modified cyanobacteria in sea water inside a clear plastic enclosure so they first make sugar (pyruvate) from CO_2 and the water via photosynthesis. Then, the bacteria secrete ethanol from the cell into the salt water. As the day progresses, and the solar radiation intensifies, ethanol concentrations build up and the ethanol itself evaporates onto the roof of the enclosure. As the sun recedes, evaporated ethanol and water condense into droplets, which run along the plastic walls and into ethanol collectors, from where it is extracted from the enclosure with the water and ethanol separated outside the enclosure. As of March 2013, Algenol was claiming to have tested its technology in Florida and to have achieved yields of 9,000 US gallons per acre per year. This could potentially meet US demands for ethanol in gasoline in 2025, assuming a B30 blend, from an area of around half the size of California's San Bernardino County, requiring less than one-tenth of the area than ethanol from other biomass, such as corn, and only very limited amounts of fresh water.

Cyanobacteria may possess the ability to produce substances that could one day serve as anti-inflammatory agents and combat bacterial infections in humans.

Spirulina's extracted blue color is used as a natural food coloring in gum and candy.

Researchers from several space agencies argue that cyanobacteria could be used for producing goods for human consumption (food, oxygen...) in future manned outposts on Mars, by transforming materials available on this planet.

Health Risks

Cyanobacteria can produce neurotoxins, cytotoxins, endotoxins, and hepatotoxins (i.e. the microcystin-producing bacteria species microcystis), and are called cyanotoxins.

Specific toxins include, anatoxin-a, anatoxin-as, aplysiatoxin, cyanopeptolin, cylindrospermopsin, domoic acid, nodularin R (from *Nodularia*), neosaxitoxin, and saxitoxin. Cyanobacteria reproduce explosively under certain conditions. This results in algal blooms, which can become harmful to other species, and pose a danger to humans and animals, if the cyanobacteria involved produce toxins. Several cases of human poisoning have been documented, but a lack of knowledge prevents an accurate assessment of the risks.

Recent studies suggest that significant exposure to high levels of cyanobacteria producing toxins such as BMAA can cause amyotrophic lateral sclerosis (ALS). People living within a half-mile of cyanobacterially contaminated lakes have had a 2.3-times greater risk of developing ALS than the rest of the population; people around New Hampshire's Lake Mascoma had an up to 25 times greater risk of ALS than the expected incidence. BMAA from desert crusts found throughout Qatar might have contributed to higher rates of ALS in Gulf War veterans.

Chemical Control

Several chemicals can eliminate blue-green algal blooms from water-based systems. They include: calcium hypochlorite, copper sulphate, cupricide, and simazine. The calcium hypochlorite amount needed varies depending on the cyanobacteria bloom, and treatment is needed periodically. According to the Department of Agriculture Australia, a rate of 12 g of 70% material in 1000 l of water is often effective to treat a bloom. Copper sulfate is also used commonly, but no longer recommended by the Australian Department of Agriculture, as it kills livestock, crustaceans, and fish. Culpricide is a chelated copper product that eliminates blooms with lower toxicity risks than copper sulfate. Dosage recommendations vary from 190 ml to 4.8 l per 1000 m². Ferric alum treatments at the rate of 50 mg/l will reduce algae blooms. Simazine, which is also a herbicide, will continue to kill blooms for several days after an application. Simazine is marketed at different strengths (25, 50, and 90%), the recommended amount needed for one cubic meter of water per product is 25% product 8 ml; 50product 4 ml; or 90% product 2.2 ml.

Dietary Supplementation

Spirulina tablets

Some cyanobacteria are sold as food, notably *Aphanizomenon flos-aquae* and *Arthrospira platensis* (Spirulina).

Despite the associated toxins which many of the members of this phylum produce, some microalgae also contain substances of high biological value, such as polyunsaturated fatty acids, amino acids (proteins), pigments, antioxidants, vitamins, and minerals. Edible blue-green algae reduce the production of pro-inflammatory cytokines by inhibiting NF-κB pathway in macrophages and splenocytes. Sulfate polysaccharides exhibit immunomodulatory, antitumor, antithrombotic, anticoagulant, anti-mutagenic, anti-inflammatory, antimicrobial, and even antiviral activity against HIV, herpes, and hepatitis.

Purple Bacteria

Purple bacteria or purple photosynthetic bacteria are proteobacteria that are phototrophic, that is, capable of producing their own food via photosynthesis. They are pigmented with bacteriochlorophyll *a* or *b*, together with various carotenoids, which give them colours ranging between purple, red, brown, and orange. They may be divided into two groups – purple sulfur bacteria (Chromatiales, in part) and purple non-sulfur bacteria (Rhodospirillaceae).

Metabolism

Photosynthesis takes place at reaction centers on the cell membrane, which is folded into the cell to form sacs, tubes, or sheets, increasing the available surface area.

Like most other photosynthetic bacteria, purple bacteria do not produce oxygen (anoxygenic), because the reducing agent (electron donor) involved in photosynthesis is not water. In some, called purple sulfur bacteria, it is either sulfide or elemental sulfur. The others, called purple non-sulfur bacteria (aka PNSB), typically use hydrogen although some may use other compounds in small amounts. At one point these were considered families, but RNA trees show the purple bacteria make up a variety of separate groups, each closer relatives of non-photosynthetic proteobacteria than one another.

The reaction centers create a charge separation through a series of favorable redox reactions, after the excitation of the special pigment pair P870. The reduction of quinones leads to the take up of 2 protons from the cytoplasm. When the quinones are eventually oxidized, they release the protons in the periplasmic side. This builds up a proton motive force that is used by ATP synthase to produce ATP from ADP and phosphate.The ATP is finally used in biosynthesis.

History

Purple bacteria were the first bacteria discovered to photosynthesize without having an oxygen byproduct. Instead, their byproduct is sulfur. This was proved by first establishing the bacteria's reactions to different concentrations of oxygen. What was found was that the bacteria moved quickly away from even the slightest trace of oxygen. Then a dish of the bacteria was taken, and a light was focused on one part of the dish leaving the rest dark. As the bacteria cannot survive without light, all the bacteria moved into the circle of light, becoming very crowded. If the bacteria's byproduct was oxygen, the distances between individuals would become larger and larger as more oxygen was produced. But because of the bacteria's behavior in the focused light, it was concluded that the bacteria's photosynthetic byproduct could not be oxygen.

Evolution

Researchers have theorized that some purple bacteria are related to the mitochondria, symbiotic bacteria in plant and animal cells today that act as organelles. Comparisons of their protein structure suggests that there is a common ancestor.

Taxonomy

Purple non-sulfur bacteria are found among the alpha and beta subgroups, including:

Rhodospirillales	
Rhodospirillaceae	e.g. *Rhodospirillum*
Acetobacteraceae	e.g. *Rhodopila*
Rhizobiales	
Bradyrhizobiaceae	e.g. *Rhodopseudomonas palustris*
Hyphomicrobiaceae	e.g. *Rhodomicrobium*
Rhodobiaceae	e.g. *Rhodobium*
Other families	
Rhodobacteraceae	e.g. *Rhodobacter*
Rhodocyclaceae	e.g. *Rhodocyclus*
Comamonadaceae	e.g. *Rhodoferax*

Purple sulfur bacteria are included among the gamma subgroup, and make up the order Chromatiales. The similarity between the photosynthetic machinery in these different lines indicates it had a common origin, either from some common ancestor or passed by lateral transfer.

Green Sulfur Bacteria

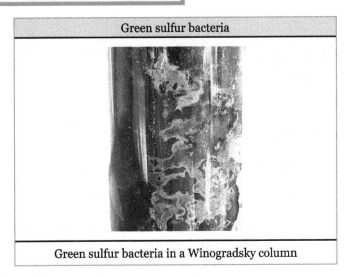

Green sulfur bacteria
Green sulfur bacteria in a Winogradsky column

The green sulfur bacteria (Chlorobiaceae) are a family of obligately anaerobic photo-autotrophic bacteria. Together with the non-photosynthetic Ignavibacteriaceae, they form the phylum Chlorobi. Most closely related to the distant Bacteroidetes, they are accordingly assigned their own phylum.

Green sulfur bacteria are nonmotile (except *Chloroherpeton thalassium*, which may glide). Photosynthesis is achieved using a Type 1 Reaction Centre using bacteriochlorophyll (BChl) *a* and in chlorosomes which employ BChl *c*, *d*, or *e*; in addition chlorophyll *a* is also present. They use sulfide ions, hydrogen or ferrous iron as an electron donor and the process is mediated by the type I reaction centre and Fenna-Matthews-Olson complex. Elemental sulfur deposited outside the cell may be further oxidized. By contrast, the photosynthesis in plants uses water as the electron donor and produces oxygen.

Chlorobium tepidum has emerged as a model organism for the group; although only 10 genomes have been sequenced, these are quite comprehensive of the family's biodiversity. Their 2-3 Mb genomes encode 1750-2800 genes, 1400-1500 of which are common to all strains. The apparent absence of two-component histidine-kinases and response regulators suggest limited phenotypic plasticity. Their small dependence on organic molecule transporters and transcription factors also indicate these organisms are adapted to a narrow range of energy-limited conditions, an ecology shared with the simpler cyanobacteria, *Prochlorococcus* and *Synechococcus*.

A species of green sulfur bacteria has been found living near a black smoker off the coast of Mexico at a depth of 2,500 m in the Pacific Ocean. At this depth, the bacterium, designated GSB1, lives off the dim glow of the thermal vent since no sunlight can penetrate to that depth.

Green sulfur bacteria appear in Lake Matano, Indonesia, at a depth of about 110–120 m. The population may include the species *Chlorobium ferrooxidans*.

Taxonomy

The currently accepted taxonomy is based on the List of Prokaryotic names with Standing in Nomenclature (LSPN)

- Phylum Chlorobi Iino et al. 2010

- Class Ignavibacteria Iino et al. 2010

  o Order Ignavibacteriales Iino et al. 2010

    ▪ Family Ignavibacteriaceae Iino et al. 2010

      ▪ Genus *Ignavibacterium* Iino et al. 2010 emend. Podosokorskaya et al. 2013

        ▪ Species *Ignavibacterium album* Iino et al. 2010 emend. Podosokorskaya et al. 2013

      ▪ Genus *Melioribacter roseus* Podosokorskaya et al. 2013

["*Melioribacter*" Podosokorskaya et al. 2011]

- Species *Melioribacter roseus* Podosokorskaya et al. 2011 ["*Melioribacter roseus*" Podosokorskaya et al. 2011]

- Class Chlorobea Cavalier-Smith 2002

  - Order Chlorobiales Gibbons and Murray 1978

    - Family Chlorobiaceae Copeland 1956

      - Genus *Ancalochloris* Gorlenko and Lebedeva 1971

        - Species *Ancalochloris perfilievii* ♪ Gorlenko and Lebedeva 1971

      - Genus *Chlorobaculum* Imhoff 2003

        - Species "*C. macestae*" ♠ Keppen et al. 2008

        - Species *C. limnaeum* Imhoff 2003

        - Species *C. parvum* Imhoff 2003

        - Species *C. tepidum* (Wahlund et al. 1996) Imhoff 2003 (type sp.) ["*Chlorobium tepidum*" Wahlund et al. 1991; *Chlorobium tepidum* Wahlund et al. 1996]

        - Species *C. thiosulfatiphilum* Imhoff 2003 ["*Chlorobium limicola* f. sp. *thiosulfatophilum*" (Larsen 1952) Pfennig & Truper 1971]

      - Genus *Chlorobium* Nadson 1906 emend. Imhoff 2003

        - Species *Chlorobium chlorovibrioides* ♦ (Gorlenko et al. 1974) Imhoff 2003

        - Species *C. bathyomarinum* ♠ Beatty et al. 2005

        - Species *C. chlorochromatii* ♠ Vogl et al. 2006 (epibiont of the phototrophic consortium *Chlorochromatium aggregatum*) ["*Chlorobium chlorochromatii*" Meschner 1957]

        - Species *C. gokarna* ♠ Anil Kumar 2005

        - Species *C. clathratiforme* (Szafer 1911) emend.

Imhoff 2003 ["*Aphanothece clathratiformis*" Szafer 1911; "*Pelodictyon lauterbornii*" Geitler 1925; *Pelodictyon clathratiforme* (Szafer 1911) Lauterborn 1913]

- Species *C. ferrooxidans* Heising et al. 1998 emend. Imhoff 2003

- Species *C. luteolum* (Schmidle 1901) emend. Imhoff 2003 ["*Aphanothece luteola*" Schmidle 1901; "*Pelodictyon aggregatum*" Perfil'ev 1914; "*Schmidlea luteola*" (Schmidle 1901) Lauterborn 1913; *Pelodictyon luteolum* (Schmidle 1901) Pfennig and Truper 1971]

- Species *C. limicola* Nadson 1906 emend. Imhoff 2003 (type sp.)

- Species *C. phaeobacteroides* Pfennig 1968 emend. Imhoff 2003

- Species *C. phaeovibrioides* Pfennig 1968 emend. Imhoff 2003

- Genus *Chloroherpeton* Gibson et al. 1985

  - Species *Chloroherpeton thalassium* Gibson et al. 1985

- Genus *Clathrochloris* Witt et al. 1989

  - Species "*Clathrochloris sulfurica*" ♠ Witt et al. 1989

- Genus *Pelodictyon* Lauterborn 1913

  - Species *Pelodictyon phaeum* Gorlenko 1972

- Genus *Prosthecochloris* Gorlenko 1970 emend. Imhoff 2003

  - Species "*P. phaeoasteroides*" ♠ Puchkova & Gorlenko 1976

  - Species "*P. indica*" ♠ Anil Kumar 2005

  - Species *P. aestuarii* Gorlenko 1970 emend. Imhoff 2003 (type sp.)

- Species *P. vibrioformis* (Pelsh 1936) Imhoff 2003 [*Chlorobium vibrioforme* Pelsh 1936]

Notes:

♪ Prokaryotes where no pure (axenic) cultures are isolated or available, i. e. not cultivated or can not be sustained in culture for more than a few serial passages

♦ Type strain lost or not available

♠ Strains found at the National Center for Biotechnology Information (NCBI) but not listed in the List of Prokaryotic names with Standing in Nomenclature (LSPN)

Heliobacteria

The heliobacteria are phototrophic: they convert light energy into chemical energy by photosynthesis and they use a Type I reaction centerHeinickel and Golbeck 2007. The primary pigment involved is bacteriochlorophyll *g*, which is unique to the group and has a unique absorption spectrum; this gives the heliobacteria their own environmental niche. Phototrophy takes place at the cell membrane, which does not form folds or compartments as it does in purple bacteria. Even though heliobacteria are phototrophic, they can grow without light by fermentation of pyruvate.

RNA trees place the heliobacteria among the Firmicutes but they do not stain gram-positively. They have no outer membrane and like certain other firmicutes (clostridia) they form heat resistant endospores, which contain high levels of calcium and dipicolinic acid. Heliobacteria are the only firmicutes known to conduct photosynthesis.

Heliobacteria are photoheterotrophic, requiring organic carbon sources, and they are exclusively anaerobic. Chlorophyll g is inactivated by the presence of oxygen, making them obligate anaerobes (they cannot survive in aerobic conditions). So far heliobacteria have only been found in soils, and are apparently widespread in the waterlogged soils of paddy fields. They are avid nitrogen fixers and are therefore probably important in the fertility of paddy fields.

Taxonomy

Heliobacteria should not be confused with *Helicobacter*, which is a different type of genus of bacteria.

Family Heliobacteriaceae

- *Candidatus* Helioclostridium♠ Girija et al. 2006

    o *Candidatus* Helioclostridium ananthapuram♠ Girija et al. 2006

- *Heliorestis* Bryantseva et al. 2000

  - *H. baculata* Bryantseva et al. 2001

  - *H. convoluta*♠ Asao et al. 2005

  - *H. daurensis* Bryantseva et al. 2000

- *Heliophilum* Ormerod et al. 1996

  - *Heliophilum fasciatum* Ormerod et al. 1996

- *Heliobacillus* Beer-Romero and Gest 1998

  - *Candidatus* H. elongatus♠ Girija et al. 2006

  - *H. mobilis* Beer-Romero and Gest 1998

- *Heliobacterium* Gest and Favinger 1985

  - *H. aridinosum*♠ Girija et al. 2006

  - *H. chlorum* Gest and Favinger 1985

  - *H. gestii* Ormerod et al. 1996

  - *H. modesticaldum* Kimble et al. 1996

  - *H. sulfidophilum* Bryantseva et al. 2001

  - *H. undosum* Bryantseva et al. 2001

References

- Hine, Robert (2005). The Facts on File dictionary of biology. Infobase Publishing. p. 175. ISBN 978-0-8160-5648-4.

- Campbell, Neil A.; Reece, Jane B.; Urry, Lisa A.; Cain, Michael L.; Wasserman, Steven A.; Minorsky, Peter V.; Jackson, Robert B. (2008). Biology (8th ed.). p. 564. ISBN 978-0-8053-6844-4.

- Nabors, Murray W. (2004). Introduction to Botany. San Francisco, CA: Pearson Education, Inc. ISBN 978-0-8053-4416-5.

- Losos, Jonathan B.; Mason, Kenneth A.; Singer, Susan R. (2007). Biology (8 ed.). McGraw-Hill. ISBN 978-0-07-304110-0.

- Silva PC, Basson PW and Moe RL (1996) Catalogue of the Benthic Marine Algae of the Indian Ocean page 2, University of California Press. ISBN 978-0-520-91581-7.

- Brodo, Irwin M; Sharnoff, Sylvia Duran; Sharnoff, Stephen; Laurie-Bourque, Susan (2001). Lichens of North America. New Haven: Yale University Press. p. 8. ISBN 978-0-300-08249-4.

- Taylor, Dennis L (1983). "The coral-algal symbiosis". In Goff, Lynda J. Algal Symbiosis: A Continuum of Interaction Strategies. CUP Archive. pp. 19–20. ISBN 978-0-521-25541-7.

- Mondragón, Jennifer; Mondragón, Jeff (2003). Seaweeds of the Pacific Coast. Monterey, California: Sea Challengers Publications. ISBN 978-0-930118-29-7.

- Simoons, Frederick J (1991). "6, Seaweeds and Other Algae". Food in China: A Cultural and Historical Inquiry. CRC Press. pp. 179–190. ISBN 978-0-936923-29-1.

- Schopf, J. W. (2012) "The fossil record of cyanobacteria", pp. 15–36 in Brian A. Whitton (Ed.) Ecology of Cyanobacteria II: Their Diversity in Space and Time. ISBN 9789400738553.

- Enrique Flores AH (2008). The Cyanobacteria: Molecular Biology, Genomics and Evolution. Horizon. p. 3. ISBN 1-904455-15-8.

- Blankenship, Robert (2009). Molecular Mechanisms of Photosynthesis. Blackwell Publishing. pp. 95–109. ISBN 978-0-632-04321-7.

7

Artificial Photosynthesis: An Overview

Bio-inspired energy technology may one day be equipped with devices that can mimic the natural photosynthesis process and convert water into hydrogen ions and oxygen through the use of sunlight. Such technologies, when perfected, can provide immediately consumable energy which is completely environment friendly. This chapter will provide an integrated understanding of artificial photosynthesis.

Artificial Photosynthesis

A sample of a photoelectric cell in a lab environment. Catalysts are added to the cell, which is submerged in water and illuminated by simulated sunlight. The bubbles seen are oxygen (forming on the front of the cell) and hydrogen (forming on the back of the cell).

Artificial photosynthesis is a chemical process that replicates the natural process of photosynthesis, a process that converts sunlight, water, and carbon dioxide into carbohydrates and oxygen. The term, artificial photosynthesis, is commonly used to refer to any scheme for capturing and storing the energy from sunlight in the chemical bonds of a fuel (a solar fuel). Photocatalytic water splitting converts water into hydrogen ions and oxygen, and is a main research area in artificial photosynthesis. Light-driven carbon dioxide reduction is another studied process, that replicates natural carbon fixation.

Research developed in this field encompasses the design and assembly of devices for the direct production of solar fuels, photoelectrochemistry and its application in fuel cells, and the engineering of enzymes and photoautotrophic microorganisms for microbial biofuel and biohydrogen production from sunlight. Many, if not most, of the artificial approaches to artificial photosynthesis are bio-inspired, i.e., they rely on biomimetics.

Overview

The photosynthetic reaction can be divided into two half-reactions of oxidation and reduction, both of which are essential to producing fuel. In plant photosynthesis, water molecules are photo-oxidized to release oxygen and protons. The second stage of plant photosynthesis (also known as the Calvin-Benson cycle) is a light-independent reaction that converts carbon dioxide into glucose (fuel). Researchers of artificial photosynthesis are developing photocatalysts that are able to perform both of these reactions. Furthermore, the protons resulting from water splitting can be used for hydrogen production. These catalysts must be able to react quickly and absorb a large percentage of the incident solar photons.

Whereas photovoltaics can provide energy directly from sunlight, the inefficiency of fuel production from photovoltaic electricity (indirect process) and the fact that sunshine is not constant throughout the day sets a limit to its use. One way of using natural photosynthesis is via the production of a biofuel, which is an indirect process that suffers from low energy conversion efficiency (due to photosynthesis' own low efficiency in converting sunlight to biomass), the cost of harvesting and transporting the fuel, and clashes with the increasing need of land mass for food production. Artificial photosynthesis aims then to produce a fuel from sunlight that can be conveniently stored and used when sunlight is not available, by using direct processes, that is, to produce a solar fuel. With the development of catalysts able to reproduce the key steps of photosynthesis, water and sunlight would ultimately be the only needed sources for clean energy production. The only by-product would be oxygen, and production of a solar fuel has the potential to be cheaper than gasoline.

One process for the creation of a clean and affordable energy supply is the development of photocatalytic water splitting under solar light. This method of sustainable hydrogen production is a key objective in the development of alternative energy systems. It is also predicted to be one of the more, if not the most, efficient ways of obtaining hydrogen from water. The conversion of solar energy into hydrogen via a water-splitting process assisted by photosemiconductor catalysts is one of the most promising technologies in development. This process has the potential for large quantities of hydrogen to be generated in an ecologically sound manner. The conversion of solar energy into a clean fuel (H_2) under ambient conditions is one of the greatest challenges facing scientists in the twenty-first century.

Two approaches are generally recognized in the construction of solar fuel cells for hydrogen production:

- A homogeneous system is one where catalysts are not compartmentalized, that is, components are present in the same compartment. This means that hydrogen and oxygen are produced in the same location. This can be a drawback, since they compose an explosive mixture, demanding gas product separation. Also, all components must be active in approximately the same conditions (e.g., pH).

- A heterogeneous system has two separate electrodes, an anode and a cathode, making possible the separation of oxygen and hydrogen production. Furthermore, different components do not necessarily need to work in the same conditions. However, the increased complexity of these systems makes them harder to develop and more expensive.

Another area of research within artificial photosynthesis is the selection and manipulation of photosynthetic microorganisms, namely green microalgae and cyanobacteria, for the production of solar fuels. Many strains are able to produce hydrogen naturally, and scientists are working to improve them. Algae biofuels such as butanol and methanol are produced both at laboratory and commercial scales. This approach has benefited from the development of synthetic biology, which is also being explored by the J. Craig Venter Institute to produce a synthetic organism capable of biofuel production.

History

The artificial photosynthesis was first anticipated by the Italian chemist Giacomo Ciamician in 1912. In a lecture that was later published in Science he proposed a switch from the use of fossil fuels to radiant energy provided by the sun and captured by technical photochemistry devices. In this switch he saw a possibility to close the gap between the rich north and poor south and ventured a guess that this switch from coal to solar energy would "not be harmful to the progress and to human happiness."

In the late 60s, Akira Fujishima discovered the photocatalytic properties of titanium dioxide, the so-called Honda-Fujishima effect, which could be used for hydrolysis.

The Swedish Consortium for Artificial Photosynthesis, the first of its kind, was established in 1994 as a collaboration between groups of three different universities, Lund, Uppsala and Stockholm, being presently active around Lund and the Ångström Laboratories in Uppsala. The consortium was built with a multidisciplinary approach to focus on learning from natural photosynthesis and applying this knowledge in biomimetic systems.

Research into artificial photosynthesis is undergoing a boom at the beginning of the 21st century. In 2000, Commonwealth Scientific and Industrial Research Organisation (CSIRO) researchers publicized their intent to focus on carbon dioxide capture and its conversion to hydrocarbons. In 2003, the Brookhaven National Laboratory announced the discovery of an important intermediate step in the reduction of CO_2 to CO (the simplest possible carbon dioxide reduction reaction), which could lead to better catalysts.

One of the drawbacks of artificial systems for water-splitting catalysts is their general reliance on scarce, expensive elements, such as ruthenium or rhenium. In 2008, with the funding of the United States Air Force Office of Scientific Research, MIT chemist and head of the Solar Revolution Project Daniel G. Nocera and postdoctoral fellow Matthew Kanan attempted to circumvent this issue by using a catalyst containing the cheaper and more abundant elements cobalt and phosphate. The catalyst was able to split water into oxygen and protons using sunlight, and could potentially be coupled to a hydrogen gas producing catalyst such as platinum. Furthermore, while the catalyst broke down during catalysis, it could self-repair. This experimental catalyst design was considered a major breakthrough in the field by many researchers.

Whereas CO is the prime reduction product of CO_2, more complex carbon compounds are usually desired. In 2008, Andrew B. Bocarsly reported the direct conversion of carbon dioxide and water to methanol using solar energy in a highly efficient photochemical cell.

While Nocera and coworkers had accomplished water splitting to oxygen and protons, a light-driven process to produce hydrogen is desirable. In 2009, the Leibniz Institute for Catalysis reported inexpensive iron carbonyl complexes able to do just that. In the same year, researchers at the University of East Anglia also used iron carbonyl compounds to achieve photoelectrochemical hydrogen production with 60% efficiency, this time using a gold electrode covered with layers of indium phosphide to which the iron complexes were linked. Both of these processes used a molecular approach, where discrete nanoparticles are responsible for catalysis.

Visible light water splitting with a one piece multijunction cell was first demonstrated and patented by William Ayers at Energy Conversion Devices in 1983. This group demonstrated water photolysis into hydrogen and oxygen, now referred to as an "artificial leaf" or "wireless solar water splitting" with a low cost, thin film amorphous silicon multijunction cell directly immersed in water. Hydrogen evolved on the front amorphous silicon surface decorated with various catalysts while oxygen evolved from the back metal substrate which also eliminated the hazard of mixed hydrogen/oxygen gas evolution. A Nafion membrane above the immersed cell provided a path for proton transport. The higher photovoltage available from the multijuction thin film cell with visible light was a major advance over previous photolysis attempts with UV sensitive single junction cells. The group's patent also lists several other semiconductor multijunction compositions in addition to amorphous silicon.

In 2009, F. del Valle and K. Domen showed the impact of the thermal treatment in a closed atmosphere using Cd1-xZnxS photocatalysts. Cd1-xZnxS solid solution reports high activity in hydrogen production from water splitting under sunlight irradiation. A mixed heterogeneous/molecular approach by researchers at the University of California, Santa Cruz, in 2010, using both nitrogen-doped and cadmium selenide quantum dots-sensitized titanium dioxide nanoparticles and nanowires, also yielded photoproduced hydrogen.

Artificial photosynthesis remained an academic field for many years. However, in the beginning of 2009, Mitsubishi Chemical Holdings was reported to be developing its own artificial photosynthesis research by using sunlight, water and carbon dioxide to "create the carbon building blocks from which resins, plastics and fibers can be synthesized." This was confirmed with the establishment of the KAITEKI Institute later that year, with carbon dioxide reduction through artificial photosynthesis as one of the main goals.

In 2010, the DOE established, as one of its Energy Innovation Hubs, the Joint Center for Artificial Photosynthesis. The mission of JCAP is to find a cost-effective method to produce fuels using only sunlight, water, and carbon-dioxide as inputs. JCAP is led by a team from Caltech, led by Professor Nathan Lewis and brings together more than 120 scientists and engineers from Caltech and its lead partner, Lawrence Berkeley National Laboratory. JCAP also draws on the expertise and capabilities of key partners from Stanford University, the University of California at Berkeley, UCSB, UCI, and UCSD, and the Stanford Linear Accelerator. In addition, JCAP serves as a central hub for other solar fuels research teams across the United States, including 20 DOE Energy Frontier Research Center. The program has a budget of $122M over five years, subject to Congressional appropriation

Also in 2010, a team led by professor David Wendell at the University of Cincinnati successfully demonstrated photosynthesis in an artificial construct consisting of enzymes suspended in a foam housing.

In 2011, Daniel Nocera and his research team announced the creation of the first practical artificial leaf. In a speech at the 241st National Meeting of the American Chemical Society, Nocera described an advanced solar cell the size of a poker card capable of splitting water into oxygen and hydrogen, approximately ten times more efficient than natural photosynthesis. The cell is mostly made of inexpensive materials that are widely available, works under simple conditions, and shows increased stability over previous catalysts: in laboratory studies, the authors demonstrated that an artificial leaf prototype could operate continuously for at least forty-five hours without a drop in activity. In May 2012, Sun Catalytix, the startup based on Nocera's research, stated that it will not be scaling up the prototype as the device offers few savings over other ways to make hydrogen from sunlight. Leading experts in the field have supported a proposal for a Global Project on Artificial Photosynthesis as a combined energy security and climate change solution. Conferences on this theme have been held at Lord Howe Island in 2011, at Chicheley Hall in the UK in 2014 and at Canberra and Lord Howe island in 2016.

Current Research

In energy terms, natural photosynthesis can be divided in three steps:

- Light-harvesting complexes in bacteria and plants capture photons and trans-
  duce them into electrons, injecting them into the photosynthetic chain.

- Proton-coupled electron transfer along several cofactors of the photosynthetic
  chain, causing local, spatial charge separation.

- Redox catalysis, which uses the aforementioned transferred electrons to oxidize
  water to dioxygen and protons; these protons can in some species be utilized for
  dihydrogen production.

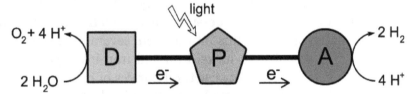

A triad assembly, with a photosensitizer (P) linked in tandem to a water oxidation catalyst (D) and a
hydrogen evolving catalyst (A). Electrons flow from D to A when catalysis occurs.

Using biomimetic approaches, artificial photosynthesis tries to construct systems doing the same type of processes. Ideally, a triad assembly could oxidize water with one
catalyst, reduce protons with another and have a photosensitizer molecule to power
the whole system. One of the simplest designs is where the photosensitizer is linked in
tandem between a water oxidation catalyst and a hydrogen evolving catalyst:

- The photosensitizer transfers electrons to the hydrogen catalyst when hit by
  light, becoming oxidized in the process.

- This drives the water splitting catalyst to donate electrons to the photosensitizer. In a triad assembly, such a catalyst is often referred to as a donor. The
  oxidized donor is able to perform water oxidation.

The state of the triad with one catalyst oxidized on one end and the second one reduced
on the other end of the triad is referred to as a charge separation, and is a driving force
for further electron transfer, and consequently catalysis, to occur. The different components may be assembled in diverse ways, such as supramolecular complexes, compartmentalized cells, or linearly, covalently linked molecules.

Research into finding catalysts that can convert water, carbon dioxide, and sunlight
to carbohydrates or hydrogen is a current, active field. By studying the natural oxygen-evolving complex (OEC), researchers have developed catalysts such as the "blue
dimer" to mimic its function or inorganic-based materials such as Birnessite with the
similar building block as the OEC. Photoelectrochemical cells that reduce carbon di-

oxide into carbon monoxide (CO), formic acid (HCOOH) and methanol (CH_3OH) are under development. However, these catalysts are still very inefficient.

Hydrogen Catalysts

Hydrogen is the simplest solar fuel to synthesize, since it involves only the transference of two electrons to two protons. It must, however, be done stepwise, with formation of an intermediate hydride anion:

$$2\,e^- + 2\,H^+ \leftrightarrow H^+ + H^- \leftrightarrow H_2$$

The proton-to-hydrogen converting catalysts present in nature are hydrogenases. These are enzymes that can either reduce protons to molecular hydrogen or oxidize hydrogen to protons and electrons. Spectroscopic and crystallographic studies spanning several decades have resulted in a good understanding of both the structure and mechanism of hydrogenase catalysis. Using this information, several molecules mimicking the structure of the active site of both nickel-iron and iron-iron hydrogenases have been synthesized. Other catalysts are not structural mimics of hydrogenase but rather functional ones. Synthesized catalysts include structural H-cluster models, a dirhodium photocatalyst, and cobalt catalysts.

Water-oxidizing Catalysts

Water oxidation is a more complex chemical reaction than proton reduction. In nature, the oxygen-evolving complex performs this reaction by accumulating reducing equivalents (electrons) in a manganese-calcium cluster within photosystem II (PS II), then delivering them to water molecules, with the resulting production of molecular oxygen and protons:

$$2\,H_2O \rightarrow O_2 + 4\,H^+ + 4e^-$$

Without a catalyst (natural or artificial), this reaction is very endothermic, requiring high temperatures (at least 2500 K).

The exact structure of the oxygen-evolving complex has been hard to determine experimentally. As of 2011, the most detailed model was from a 1.9 Å resolution crystal structure of photosystem II. The complex is a cluster containing four manganese and one calcium ions, but the exact location and mechanism of water oxidation within the cluster is unknown. Nevertheless, bio-inspired manganese and manganese-calcium complexes have been synthesized, such as $[Mn_4O_4]$ cubane-type clusters, some with catalytic activity.

Some ruthenium complexes, such as the dinuclear μ-oxo-bridged "blue dimer" (the first of its kind to be synthesized), are capable of light-driven water oxidation, thanks to being able to form high valence states. In this case, the ruthenium complex acts as both photosensitizer and catalyst.

Many metal oxides have been found to have water oxidation catalytic activity, including ruthenium(IV) oxide (RuO_2), iridium(IV) oxide (IrO_2), cobalt oxides (including nickel-doped Co_3O_4), manganese oxide (including layered MnO_2 (birnessite), Mn_2O_3), and a mix of Mn_2O_3 with $CaMn_2O_4$. Oxides are easier to obtain than molecular catalysts, especially those from relatively abundant transition metals (cobalt and manganese), but suffer from low turnover frequency and slow electron transfer properties, and their mechanism of action is hard to decipher and, therefore, to adjust.

Recently Metal-Organic Framework (MOF)-based materials have been shown to be a highly promising candidate for water oxidation with first row transition metals. The stability and tunability of this system is projected to be highly beneficial for future development.

Photosensitizers

Structure of $[Ru(bipy)_3]^{2+}$, a broadly used photosensitizer.

Nature uses pigments, mainly chlorophylls, to absorb a broad part of the visible spectrum. Artificial systems can use either one type of pigment with a broad absorption range or combine several pigments for the same purpose.

Ruthenium polypyridine complexes, in particular tris(bipyridine)ruthenium(II) and its derivatives, have been extensively used in hydrogen photoproduction due to their efficient visible light absorption and long-lived consequent metal-to-ligand charge transfer excited state, which makes the complexes strong reducing agents. Other noble metal-containing complexes used include ones with platinum, rhodium and iridium.

Metal-free organic complexes have also been successfully employed as photosensitizers. Examples include eosin Y and rose bengal. Pyrrole rings such as porphyrins have also been used in coating nanomaterials or semiconductors for both homogeneous and heterogeneous catalysis.

As part of current research efforts artificial photonic antenna systems are being studied to determine efficient and sustainable ways to collect light for artificial photosynthesis. Gion Calzaferri (2009) describes one such antenna that uses zeolite L as a host for organic dyes, to mimic plant's light collecting systems. The material may be interfaced to an external device via a stopcock intermediate.

Carbon Dioxide Reduction Catalysts

In nature, carbon fixation is done by green plants using the enzyme RuBisCO as a part of the Calvin cycle. RuBisCO is a rather slow catalyst compared to the vast majority of other enzymes, incorporating only a few molecules of carbon dioxide into ribulose-1,5-bisphosphate per minute, but does so at atmospheric pressure and in mild, biological conditions. The resulting product is further reduced and eventually used in the synthesis of glucose, which in turn is a precursor to more complex carbohydrates, such as cellulose and starch. The process consumes energy in the form of ATP and NADPH.

Artificial CO_2 reduction for fuel production aims mostly at producing reduced carbon compounds from atmospheric CO_2. Some transition metal polyphosphine complexes have been developed for this end; however, they usually require previous concentration of CO_2 before use, and carriers (molecules that would fixate CO_2) that are both stable in aerobic conditions and able to concentrate CO_2 at atmospheric concentrations haven't been yet developed. The simplest product from CO_2 reduction is carbon monoxide (CO), but for fuel development, further reduction is needed, and a key step also needing further development is the transfer of hydride anions to CO.

Other Materials and Components

Charge separation is a key property of dyad and triad assemblies. Some nanomaterials employed are fullerenes (such as carbon nanotubes), a strategy that explores the pi-bonding properties of these materials. Diverse modifications (covalent and non-covalent) of carbon nanotubes have been attempted to increase the efficiency of charge separation, including the addition of ferrocene and pyrrole-like molecules such as porphyrins and phthalocyanines.

Since photodamage is usually a consequence in many of the tested systems after a period of exposure to light, bio-inspired photoprotectants have been tested, such as carotenoids (which are used in photosynthesis as natural protectants).

Light-driven Methodologies Under Development

Photoelectrochemical Cells

Photoelectrochemical cells are a heterogeneous system that use light to produce either electricity or hydrogen. The vast majority of photoelectrochemical cells use semiconductors as catalysts. There have been attempts to use synthetic manganese com-plex-impregnated Nafion as a working electrode, but it has been since shown that the catalytically active species is actually the broken-down complex.

A promising, emerging type of solar cell is the dye-sensitized solar cell. This type of cell still depends on a semiconductor (such as TiO_2) for current conduction on one electrode, but with a coating of an organic or inorganic dye that acts as a photosensi-tizer; the counter electrode is a platinum catalyst for H_2 production. These cells have a self-repair mechanism and solar-to-electricity conversion efficiencies rivaling those of solid-state semiconductor ones.

Photocatalytic Water Splitting in Homogeneous Systems

Direct water oxidation by photocatalysts is a more efficient usage of solar energy than photoelectrochemical water splitting because it avoids an intermediate thermal or elec-trical energy conversion step.

Bio-inspired manganese clusters have been shown to possess water oxidation activity when adsorbed on clays together with ruthenium photosensitizers, although with low turnover numbers.

As mentioned above, some ruthenium complexes are able to oxidize water under solar light irradiation. Although their photostability is still an issue, many can be reactivated by a simple adjustment of the conditions they work in. Improvement of catalyst stabil-ity has been tried resorting to polyoxometalates, in particular ruthenium-based ones.

Whereas a fully functional artificial system is usually envisioned when constructing a water splitting device, some mixed approaches have been tried. One of these involve the use of a gold electrode to which photosystem II is linked; an electric current is de-tected upon illumination.

Hydrogen-producing Artificial Systems

A H-cluster FeFe hydrogenase model compound covalently linked to a ruthenium photosensitizer. The ruthenium complex absorbs light and transduces its energy to the iron compound, which can then reduce protons to H_2.

The simplest photocatalytic hydrogen production unit consists of a hydrogen-evolving catalyst linked to a photosensitizer. In this dyad assembly, a so-called sacrificial donor for the photosensitizer is needed, that is, one that is externally supplied and replenished; the photosensitizer donates the necessary reducing equivalents to the hydrogen-evolving catalyst, which uses protons from a solution where it is immersed or dissolved in. Cobalt compounds such as cobaloximes are some of the best hydrogen catalysts, having been coupled to both metal-containing and metal-free photosensitizers. The first H-cluster models linked to photosensitizers (mostly ruthenium photosensitizers, but also porphyrin-derived ones) were prepared in the early 2000s. Both types of assembly are under development to improve their stability and increase their turnover numbers, both necessary for constructing a sturdy, long-lived solar fuel cell.

As with water oxidation catalysis, not only fully artificial systems have been idealized: hydrogenase enzymes themselves have been engineered for photoproduction of hydrogen, by coupling the enzyme to an artificial photosensitizer, such as $[Ru(bipy)_3]^{2+}$ or even photosystem I.

NADP⁺/NADPH Coenzyme-inspired Catalyst

In natural photosynthesis, the $NADP^+$ coenzyme is reducible to NADPH through binding of a proton and two electrons. This reduced form can then deliver the proton and electrons, potentially as a hydride, to reactions that culminate in the production of carbohydrates (the Calvin cycle). The coenzyme is recyclable in a natural photosynthetic cycle, but this process is yet to be artificially replicated.

A current goal is to obtain an NADPH-inspired catalyst capable of recreating the natural cyclic process. Utilizing light, hydride donors would be regenerated and produced where the molecules are continuously used in a closed cycle. Brookhaven chemists are now using a ruthenium-based complex to serve as the acting model. The complex is proven to perform correspondingly with NADP+/NADPH, behaving as the foundation for the proton and two electrons needed to convert acetone to isopropanol.

Currently, Brookhaven researchers are aiming to find ways for light to generate the hydride donors. The general idea is to use this process to produce fuels from carbon dioxide.

Photobiological Production of Fuels

Some photoautotrophic microorganisms can, under certain conditions, produce hydrogen. Nitrogen-fixing microorganisms, such as filamentous cyanobacteria, possess the enzyme nitrogenase, responsible for conversion of atmospheric N_2 into ammonia; molecular hydrogen is a byproduct of this reaction, and is many times not released by the microorganism, but rather taken up by a hydrogen-oxidizing (uptake) hydrogenase. One way of forcing these organisms to produce hydrogen is then to annihilate uptake hydrogenase activity. This has been done on a strain of *Nostoc punctiforme*: one of the structural genes of the NiFe uptake hydrogenase was inactivated by insertional mutagenesis, and the mutant strain showed hydrogen evolution under illumination.

Many of these photoautotrophs also have bidirectional hydrogenases, which can produce hydrogen under certain conditions. However, other energy-demanding metabolic pathways can compete with the necessary electrons for proton reduction, decreasing the efficiency of the overall process; also, these hydrogenases are very sensitive to oxygen.

Several carbon-based biofuels have also been produced using cyanobacteria, such as 1-butanol.

Synthetic biology techniques are predicted to be useful in this field. Microbiological and enzymatic engineering have the potential of improving enzyme efficiency and robustness, as well as constructing new biofuel-producing metabolic pathways in photoautotrophs that previously lack them, or improving on the existing ones. Another field under development is the optimization of photobioreactors for commercial application.

Employed Research Techniques

Research in artificial photosynthesis is necessarily a multidisciplinary field, requiring a multitude of different expertise. Some techniques employed in making and investigating catalysts and solar cells include:

- Organic and inorganic chemical synthesis.

- Electrochemistry methods, such as photoelectrochemistry, cyclic voltammetry, electrochemical impedance spectroscopy Dielectric spectroscopy, and bulk electrolysis.

- Spectroscopic methods:

    o fast techniques, such as time-resolved spectroscopy and ultrafast laser spectroscopy;

    o magnetic resonance spectroscopies, such as nuclear magnetic resonance, electron paramagnetic resonance;

    o X-ray spectroscopy methods, including x-ray absorption such as XANES and EXAFS, but also x-ray emission.

- Crystallography.

- Molecular biology, microbiology and synthetic biology methodologies.

Advantages, Disadvantages, and Efficiency

Advantages of solar fuel production through artificial photosynthesis include:

- The solar energy can be immediately converted and stored. In photovoltaic cells, sunlight is converted into electricity and then converted again into chemical energy for storage, with some necessary loss of energy associated with the second conversion.

- The byproducts of these reactions are environmentally friendly. Artificially photosynthesized fuel would be a carbon-neutral source of energy, which could be used for transportation or homes.

Disadvantages include:

- Materials used for artificial photosynthesis often corrode in water, so they may be less stable than photovoltaics over long periods of time. Most hydrogen catalysts are very sensitive to oxygen, being inactivated or degraded in its presence; also, photodamage may occur over time.

- The overall cost is not yet advantageous enough to compete with fossil fuels as a commercially viable source of energy.

A concern usually addressed in catalyst design is efficiency, in particular how much of the incident light can be used in a system in practice. This is comparable with photosynthetic efficiency, where light-to-chemical-energy conversion is measured. Photosynthetic organisms are able to collect about 50% of incident solar radiation, however the theoretical limit of photosynthetic efficiency is 4.6 and 6.0% for C3 and C4 plants respectively. In reality, the efficiency of photosynthesis is much lower and is usually below 1%, with some exceptions such as sugarcane in tropical climate. In contrast, the highest reported efficiency for artificial photosynthesis lab prototypes is 22.4%. However, plants are efficient in using CO_2 at atmospheric concentrations, something that artificial catalysts still cannot perform.

Photocatalytic Water Splitting

Photocatalytic water splitting is an artificial photosynthesis process with photocatalysis in a photoelectrochemical cell used for the dissociation of water into its constituent parts, hydrogen (H2) and oxygen (O2), using either artificial or natural light. Theoretically, only solar energy (photons), water, and a catalyst are needed. This topic is the focus of much research, but thus far no technology has been commercialized

Hydrogen fuel production has gained increased attention as oil and other nonrenewable fuels become increasingly depleted and expensive. Methods such as photocatalytic water splitting are being investigated to produce hydrogen, a clean-burning fuel. Water splitting holds particular promise since it utilizes water, an inexpensive renewable resource. Photocatalytic water splitting has the simplicity of using a powder in solution and sunlight to produce H2 and O2 from water and can provide a clean, renewable energy, without producing greenhouse gases or having many adverse effects on the atmosphere.

Concepts

When H2O is split into O2 and H2, the stoichiometric ratio of its products is 2:1:

$$2\,H_2O \quad \overset{photon\,energy\,>1.23eV}{\rightleftharpoons} \quad 2\,H_2 + O_2$$

The process of water-splitting is a highly endothermic process ($\Delta H > 0$). Water splitting occurs naturally in photosynthesis when photon energy is absorbed and converted into the chemical energy through a complex biological pathway. However, production of hydrogen from water requires large amounts of input energy, making it incompatible with existing energy generation. For this reason, most commercially produced hydrogen gas is produced from natural gas.

There are several strict requirements for a photocatalyst to be useful for water splitting. The minimum potential difference (voltage) needed to split water is 1.23V at 0 pH. Since the minimum band gap for successful water splitting at pH=0 is 1.23 eV, corresponding to light of 1008 nm, the electrochemical requirements can theoretically reach down into infrared light, albeit with negligible catalytic activity. These values are true only for a completely reversible reaction at standard temperature and pressure (1 bar and 25 °C).

Theoretically, infrared light has enough energy to split water into hydrogen and oxygen; however, this reaction is kinetically very slow because the wavelength is greater than 380 nm. The potential must be less than 3.0V to make efficient use of the energy present across the full spectrum of sunlight. Water splitting can transfer charges, but not be able to avoid corrosion for long term stability. Defects within crystalline photocatalysts can act as recombination sites, ultimately lowering efficiency.

Under normal conditions due to the transparency of water to visible light photolysis can only occur with a radiation wavelength of 180 nm or shorter. We see then that, assuming a perfect system, the minimum energy input is 6.893 eV.

Materials used in photocatalytic water splitting fulfill the band requirements outlined previously and typically have dopants and/or co-catalysts added to optimize their performance. A sample semiconductor with the proper band structure is titanium dioxide (TiO2). However, due to the relatively positive conduction band of TiO2, there is little driving force for H2 production, so TiO2 is typically used with a co-catalyst such as platinum (Pt) to increase the rate of H2 production. It is routine to add co-catalysts to spur H2 evolution in most photocatalysts due to the conduction band placement. Most semiconductors with suitable band structures to split water absorb mostly UV light; in order to absorb visible light, it is necessary to narrow the band gap. Since the conduction band is fairly close to the reference potential for H2 formation, it is preferable to alter the valence band to move it closer to the potential for O2 formation, since there is a greater natural overpotential.

Photocatalysts can suffer from catalyst decay and recombination under operating conditions. Catalyst decay becomes a problem when using a sulfide-based photocatalyst such as cadmium sulfide (CdS), as the sulfide in the catalyst is oxidized to elemental sulfur at the same potentials used to split water. Thus, sulfide-based photocatalysts are not viable without sacrificial reagents such as sodium sulfide to replenish any sulfur lost, which effectively changes the main reaction to one of hydrogen evolution as opposed to water splitting. Recombination of the electron-hole pairs needed for photocatalysis can occur with any catalyst and is dependent on the defects and surface area of the catalyst; thus, a high degree of crystallinity is required to avoid recombination at the defects.

The conversion of solar energy to hydrogen by means of photocatalysis is one of the most interesting ways to achieve clean and renewable energy systems. However, if this process is assisted by photocatalysts suspended directly in water instead of using a photovoltaic and electrolytic system the reaction is in just one step, and can therefore be more efficient.

Method of Evaluation

Photocatalysts must conform to several key principles in order to be considered effective at water splitting. A key principle is that H2 and O2 evolution should occur in a stoichiometric 2:1 ratio; significant deviation could be due to a flaw in the experimental setup and/or a side reaction, both of which do not indicate a reliable photocatalyst for water splitting. The prime measure of photocatalyst effectiveness is quantum yield (QY), which is:

$$QY \text{ (\%)} = \text{(Photochemical reaction rate)} / \text{(Photon absorption rate)} \times 100\%$$

This quantity is a reliable determination of how effective a photocatalyst is; however, it can be misleading due to varying experimental conditions. To assist in comparison, the rate of gas evolution can also be used; this method is more problematic on its own because it is not normalized, but it can be useful for a rough comparison and is consistently reported in the literature. Overall, the best photocatalyst has a high quantum yield and gives a high rate of gas evolution.

The other important factor for a photocatalyst is the range of light absorbed; though UV-based photocatalysts will perform better per photon than visible light-based photocatalysts due to the higher photon energy, far more visible light reaches the Earth's surface than UV light. Thus, a less efficient photocatalyst that absorbs visible light may ultimately be more useful than a more efficient photocatalyst absorbing solely light with smaller wavelengths.

Photocatalyst Systems

Cd1-xZnxS

Solid solutions Cd1-xZnxS with different Zn concentration (0.2 < x < 0.35) has been investigated in the production of hydrogen from aqueous solutions containing SO32−/S2− as sacrificial reagents under visible light. Textural, structural and surface catalyst properties were determinedby N2 adsorption isotherms, UV–vis spectroscopy, SEM and XRD and related to the activity results in hydrogen production from water splitting under visible light irradiation. It was found that the crystallinity and energy band structure of the Cd1-xZnxS solid solutions depend on their Zn atomic concentration. The hydrogen production rate was found to increase gradually when the Zn concentration on photocatalysts increases from 0.2 to 0.3. Subsequent increase in the Zn fraction up to 0.35 leads to lower hydrogen production. Variation in photoactivity is analyzed in terms of changes in crystallinity, level of conduction band and light absorption ability of Cd1-xZnxS solid solutions derived from their Zn atomic concentration.

NaTaO3:La

NaTaO3:La yields the highest water splitting rate of photocatalysts without using sacrificial reagents. This UV-based photocatalyst was shown to be highly effective with water splitting rates of 9.7 mmol/h and a quantum yield of 56%. The nanostep structure of the material promotes water splitting as edges functioned as H2 production sites and the grooves functioned as O2 production sites. Addition of NiO particles as cocatalysts assisted in H2 production; this step was done by using an impregnation method with an aqueous solution of Ni(NO3)2•6H2O and evaporating the solution in the presence of the photocatalyst. NaTaO3 has a conduction band higher than that of NiO, so photo-generated electrons are more easily transferred to the conduction band of NiO for H2 evolution.

K3Ta3B2O12

K3Ta3B2O12, another catalyst activated by solely UV light and above, does not have the performance or quantum yield of NaTaO3:La. However, it does have the ability to split water without the assistance of cocatalysts and gives a quantum yield of 6.5% along with a water splitting rate of 1.21 mmol/h. This ability is due to the pillared structure of the photocatalyst, which involves TaO6 pillars connected by BO3 triangle units. Loading with NiO did not assist the photocatalyst due to the highly active H2 evolution sites.

(Ga.82Zn.18)(N.82O.18)

(Ga.82Zn.18)(N.82O.18) has the highest quantum yield in visible light for visible light-based photocatalysts that do not utilize sacrificial reagents as of October 2008. The photocatalyst gives a quantum yield of 5.9% along with a water splitting rate of 0.4 mmol/h. Tuning the catalyst was done by increasing calcination temperatures for the final step in synthesizing the catalyst. Temperatures up to 600 °C helped to reduce the number of defects, though temperatures above 700 °C destroyed the local structure around zinc atoms and was thus undesirable. The treatment ultimately reduced the amount of surface Zn and O defects, which normally function as recombination sites, thus limiting photocatalytic activity. The catalyst was then loaded with Rh2-yCryO3 at a rate of 2.5 wt % Rh and 2 wt% Cr to yield the best performance.

Cobalt Based Systems

Photocatalysts based on cobalt have been reported. Members are tris(bipyridine) cobalt(II), compounds of cobalt ligated to certain cyclic polyamines, and certain cobaloximes.

In 2014 researchers announced an approach that connected a chromophore to part of a larger organic ring that surrounded a cobalt atom. The process is less efficient than using a platinum catalyst, cobalt is less expensive, potentially reducing total costs. The process uses one of two supramolecular assemblies based on Co(II)-templated coordination of Ru(bpy)$_3$2+ (bpy = 2,2'-bipyridyl) analogues as photosensitizers and electron donors to a cobaloxime macrocycle. The Co(II) centres of both assemblies are high spin, in contrast to most previously described cobaloximes. Transient absorption optical spectroscopies include that charge recombination occurs through multiple ligand states present within the photosensitizer modules.

Bismuth Vanadate

Bismuth based systems have been demonstrated to have an efficiency of 5% with the advantage of a very simple and cheap catalyst.

Tungsten Diselenide (WSe$_2$)

Tungsten diselenide may have a role in future hydrogen fuel production, as a recent discovery in 2015 by scientists in Switzerland revealed that the compound's own photocatalytic properties might be a key to significantly more efficient electrolysis of water to produce hydrogen fuel.

III-V Semiconductor Systems

Systems based on the material class of III-V semiconductors, such as InGaP, enable currently the highest solar-to-hydrogen efficiencies of up to 14%. Long-term stability of these high-cost high-efficiency systems does, however, remain an issue.

References

- "Discovery Brightens Solar's Future, Energy Costs to Be Cut". nbcnews.com. NBC News from Reuters. July 2, 2015. Retrieved July 2, 2015.

- "A less-expensive way to duplicate the complicated steps of photosynthesis in making fuel". KurzweilAI. doi:10.1039/C3CP54420F. Retrieved 2014-01-23.

- "Hydrogen from Water in a Novel Recombinant Cyanobacterial System". J. Craig Venter Institute. Retrieved 17 January 2012.

- Lachance, Molly. "AF Funding Enables Artificial Photosynthesis". Wright-Patterson Air Force Base News. Wright-Patterson Air Force Base. Retrieved 19 January 2012.

- Trafton, Anne. "'Major discovery' from MIT primed to unleash solar revolution". MIT News. Massachusetts Institute of Technology. Retrieved 10 January 2012.

- Kleiner, Kurt. "Electrode lights the way to artificial photosynthesis". NewScientist. Reed Business Information Ltd. Retrieved 10 January 2012.

8

Evolution of Photosynthesis

The first photosynthetic organisms can be traced back to 2450 million years ago according to geological records. This chapter chronicles the evolutionary history of photosynthetic organisms and photosynthetic pathways. The reader is presented with information about the families of phototrophs and other photosynthetic organisms that exist on Earth. It provides a comprehensive overview of the evolution of photosynthesis.

The first photosynthetic organisms probably evolved early in the evolutionary history of life and most likely used reducing agents such as hydrogen or hydrogen sulfide as sources of electrons, rather than water. There are three major metabolic pathways by which photosynthesis is carried out: C_3 photosynthesis, C_4 photosynthesis, and CAM photosynthesis. C_3 photosynthesis is the oldest and most common form.

The biochemical capacity to use water as the source for electrons in photosynthesis evolved once, in a common ancestor of extant cyanobacteria. The geological record indicates that this transforming event took place early in Earth's history, at least 2450–2320 million years ago (Ma), and, it is speculated, much earlier. Available evidence from geobiological studies of Archean (>2500 Ma) sedimentary rocks indicates that life existed 3500 Ma, but the question of when oxygenic photosynthesis evolved is still unanswered. A clear paleontological window on cyanobacterial evolution opened about 2000 Ma, revealing an already-diverse biota of blue-greens. Cyanobacteria remained principal primary producers throughout the Proterozoic Eon (2500–543 Ma), in part because the redox structure of the oceans favored photoautotrophs capable of nitrogen fixation. Green algae joined blue-greens as major primary producers on continental shelves near the end of the Proterozoic, but only with the Mesozoic (251–65 Ma) radiations of dinoflagellates, coccolithophorids, and diatoms did primary production in marine shelf waters take modern form. Cyanobacteria remain critical to marine ecosystems as primary producers in oceanic gyres, as agents of biological nitrogen fixation, and, in modified form, as the plastids of marine algae.

Early photosynthetic systems, such as those from green and purple sulfur and green and purple nonsulfur bacteria, are thought to have been anoxygenic, using various molecules as electron donors. Green and purple sulfur bacteria are thought to have used hydrogen and sulfur as an electron donor. Green nonsulfur bacteria used various amino and other organic acids. Purple nonsulfur bacteria used a variety of nonspecific organic molecules.

Fossils of what are thought to be filamentous photosynthetic organisms have been dated at 3.4 billion years old.

The main source of oxygen in the atmosphere is oxygenic photosynthesis, and its first appearance is sometimes referred to as the oxygen catastrophe. Geological evidence suggests that oxygenic photosynthesis, such as that in cyanobacteria, became important during the Paleoproterozoic era around 2 billion years ago. Modern photosynthesis in plants and most photosynthetic prokaryotes is oxygenic. Oxygenic photosynthesis uses water as an electron donor, which is oxidized to molecular oxygen (O2) in the photosynthetic reaction center.

Symbiosis and the Origin of Chloroplasts

Several groups of animals have formed symbiotic relationships with photosynthetic algae. These are most common in corals, sponges and sea anemones. It is presumed that this is due to the particularly simple body plans and large surface areas of these animals compared to their volumes. In addition, a few marine mollusks *Elysia viridis* and *Elysia chlorotica* also maintain a symbiotic relationship with chloroplasts they capture from the algae in their diet and then store in their bodies. This allows the mollusks to survive solely by photosynthesis for several months at a time. Some of the genes from the plant cell nucleus have even been transferred to the slugs, so that the chloroplasts can be supplied with proteins that they need to survive.

An even closer form of symbiosis may explain the origin of chloroplasts. Chloroplasts have many similarities with photosynthetic bacteria, including a circular chromosome, prokaryotic-type ribosomes, and similar proteins in the photosynthetic reaction center. The endosymbiotic theory suggests that photosynthetic bacteria were acquired (by endocytosis) by early eukaryotic cells to form the first plant cells. Therefore, chloroplasts may be photosynthetic bacteria that adapted to life inside plant cells. Like mitochondria, chloroplasts still possess their own DNA, separate from the nuclear DNA of their plant host cells and the genes in this chloroplast DNA resemble those in cyanobacteria. DNA in chloroplasts codes for redox proteins such as photosynthetic reaction centers. The CoRR Hypothesis proposes that this **Co**-location is required for Redox Regulation.

A 2010 study by researchers at Tel Aviv University discovered that the Oriental hornet (*Vespa orientalis*) converts sunlight into electric power using a pigment called xanthopterin. This is the first scientific evidence of a member of the animal kingdom engaging in photosynthesis.

Evolution of Photosynthetic Pathways

Photosynthesis is not quite as simple as adding water to CO_2 to produce sugars and oxygen. A complex chemical pathway is involved, facilitated along the way by a range of enzymes and co-enzymes. The enzyme RuBisCO is responsible for "fixing" CO_2 – that is, it attaches it to a carbon-based molecule to form a sugar, which can be used by the plant, releasing an oxygen molecule along the way. However, the enzyme is notoriously inefficient, and just as effectively will also fix oxygen instead of CO_2 in a process called photorespiration. This is energetically costly as the plant has to use energy to turn the products of photorespiration back into a form that can react with CO_2.

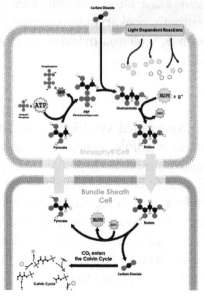

The C_4 carbon concentrating mechanism

Concentrating Carbon

The C_4 metabolic pathway is a valuable recent evolutionary innovation in plants, involving a complex set of adaptive changes to physiology and gene expression patterns. About 7600 species of plants use C_4 carbon fixation, which represents about 3% of all terrestrial species of plants. All these 7600 species are angiosperms.

C_4 plants evolved carbon concentrating mechanisms. These work by increasing the concentration of CO_2 around RuBisCO, thereby facilitating photosynthesis and decreasing photorespiration. The process of concentrating CO_2 around RuBisCO requires more energy than allowing gases to diffuse, but under certain conditions – i.e. warm temperatures (>25 °C), low CO_2 concentrations, or high oxygen concentrations – pays off in terms of the decreased loss of sugars through photorespiration.

One type of C_4 metabolism employs a so-called Kranz anatomy. This transports CO_2 through an outer mesophyll layer, via a range of organic molecules, to the central bundle sheath cells, where the CO_2 is released. In this way, CO_2 is concentrated near the site of RuBisCO operation. Because RuBisCO is operating in an environment with much more CO_2 than it otherwise would be, it performs more efficiently.

A second mechanism, CAM photosynthesis, is a carbon fixation pathway that evolved in some plants as an adaptation to arid conditions. The most important benefit of CAM to the plant is the ability to leave most leaf stomata closed during the day. This reduces water loss due to evapotranspiration. The stomata open at night to collect CO_2, which is stored as the four-carbon acid malate, and then used during photosynthesis during the day. The pre-collected CO_2 is concentrated around the enzyme RuBisCO, increasing photosynthetic efficiency. More CO_2 is then harvested from the atmosphere when stomata open, during the cool, moist nights, reducing water loss.

CAM has evolved convergently many times. It occurs in 16,000 species (about 7% of plants), belonging to over 300 genera and around 40 families, but this is thought to be a considerable underestimate. It is found in quillworts (relatives of club mosses), in ferns, and in gymnosperms, but the great majority of plants using CAM are angiosperms (flowering plants).

Evolutionary Record

These two pathways, with the same effect on RuBisCO, evolved a number of times independently – indeed, C_4 alone arose 62 times in 18 different plant families. A number of 'pre-adaptations' seem to have paved the way for C4, leading to its clustering in certain clades: it has most frequently been innovated in plants that already had features such as extensive vascular bundle sheath tissue. Many potential evolutionary pathways resulting in the C_4 phenotype are possible and have been characterised using Bayesian inference, confirming that non-photosynthetic adaptations often provide evolutionary stepping stones for the further evolution of C_4.

CAM is named after the family Crassulaceae, to which the jade plant belongs. Another example of a CAM plant is the pineapple.

The C_4 construction is most famously used by a subset of grasses, while CAM is employed by many succulents and cacti. The trait appears to have emerged during the Oligocene, around 25 to 32 million years ago; however, they did not become ecologically significant until the Miocene, 6 to 7 million years ago. Remarkably, some charcoalified fossils preserve tissue organised into the Kranz anatomy, with intact bundle sheath cells, allowing the presence C_4 metabolism to be identified without doubt at this time. Isotopic markers are used to deduce their distribution and significance.

C_3 plants preferentially use the lighter of two isotopes of carbon in the atmosphere, ^{12}C, which is more readily involved in the chemical pathways involved in its fixation. Because C_4 metabolism involves a further chemical step, this effect is accentuated. Plant material can be analysed to deduce the ratio of the heavier ^{13}C to ^{12}C. This ratio is denoted $\delta^{13}C$. C_3 plants are on average around 14‰ (parts per thousand) lighter than the atmospheric ratio, while C_4 plants are about 28‰ lighter. The $\delta^{13}C$ of CAM plants depends on the percentage of carbon fixed at night relative to what is fixed in the day, being closer to C_3 plants if they fix most carbon in the day and closer to C_4 plants if they fix all their carbon at night.

It is troublesome procuring original fossil material in sufficient quantity to analyse the grass itself, but fortunately there is a good proxy: horses. Horses were globally widespread in the period of interest, and browsed almost exclusively on grasses. There's an old phrase in isotope palæontology, "you are what you eat (plus a little bit)" – this refers to the fact that organisms reflect the isotopic composition of whatever they eat, plus a small adjustment factor. There is a good record of horse teeth throughout the globe, and their $\delta^{13}C$ has been measured. The record shows a sharp negative inflection around 6 to 7 million years ago, during the Messinian, and this is interpreted as the rise of C_4 plants on a global scale.

When is C_4 an Advantage?

While C_4 enhances the efficiency of RuBisCO, the concentration of carbon is highly energy intensive. This means that C_4 plants only have an advantage over C_3 organisms in certain conditions: namely, high temperatures and low rainfall. C_4 plants also need high levels of sunlight to thrive. Models suggest that, without wildfires removing shade-casting trees and shrubs, there would be no space for C_4 plants. But, wildfires have occurred for 400 million years – why did C_4 take so long to arise, and then appear independently so many times? The Carboniferous period (~300 million years ago) had notoriously high oxygen levels – almost enough to allow spontaneous combustion – and very low CO_2, but there is no C_4 isotopic signature to be found. And there doesn't seem to be a sudden trigger for the Miocene rise.

During the Miocene, the atmosphere and climate were relatively stable. If anything, CO_2 increased gradually from 14 to 9 million years ago before settling down to concentrations similar to the Holocene. This suggests that it did not have a key role in invoking C_4 evolution. Grasses themselves (the group which would give rise to the most occurrences of C_4) had probably been around for 60 million years or more, so had had plenty of time to evolve C_4, which, in any case, is present in a diverse range of groups and thus evolved independently. There is a strong signal of climate change in South Asia; increasing aridity – hence increasing fire frequency and intensity – may have led to an increase in the importance of grasslands. However, this is difficult to reconcile with the North American record. It is possible that the signal is entirely biological, forced by the fire- (and elephant?)- driven acceleration of grass evolution – which, both by increasing weathering and incorporating more carbon into sediments, reduced atmospheric CO_2 levels. Finally, there is evidence that the onset of C_4 from 9 to 7 million years ago is a biased signal, which only holds true for North America, from where most samples originate; emerging evidence suggests that grasslands evolved to a dominant state at least 15Ma earlier in South America.

References

- Herrero A, Flores E (2008). The Cyanobacteria: Molecular Biology, Genomics and Evolution (1st ed.). Caister Academic Press. ISBN 978-1-904455-15-8.

- C.Michael Hogan. 2011. Respiration. Encyclopedia of Earth. Eds. Mark McGinley & C.J.cleveland. National council for Science and the Environment. Washington DC

Permissions

Index